Management of the Effects of Coastal Storms

FOCUS SERIES

Series Editor André Mariotti

Management of the Effects of Coastal Storms

Policy, Scientific and Historical Perspectives

Philippe Quevauviller
Paolo Ciavola
Emmanuel Garnier

iSTE

WILEY

First published 2017 in Great Britain and the United States by ISTE Ltd and John Wiley & Sons, Inc.

ISTE Ltd
27-37 St George's Road
London SW19 4EU
UK

www.iste.co.uk

John Wiley & Sons, Inc.
111 River Street
Hoboken, NJ 07030
USA

www.wiley.com

Library of Congress Control Number: 2016961109

British Library Cataloguing-in-Publication Data
A CIP record for this book is available from the British Library
ISSN 2051-2481 (Print)
ISSN 2051-249X (Online)
ISBN 978-1-84821-762-1

Cover photo: New Brighton (UK) by Sally Priest

Contents

Preface

During the night of February 28th, 2010, the French Atlantic coastline was struck by a particularly violent storm, associated with a sea surge, which caused the death of 47 people. The flooded zones where the victims were most numerous corresponded mainly territories urbanized during the last three decades. For the littoral's population, the civil protection and the policymakers, this was a total surprise and the catastrophe was presented like a completely new and thus unforeseeable phenomenon. However, neighbouring countries close to France had coped with several similar events, sometimes more disastrous still, within the last 50 years. An example is the famous "Great storm" of 1953 which caused the death of several thousands of people in Great Britain, Belgium, the Netherlands, Germany, and which had also struck the North of France, illustrating that in France the memory of these extreme events had apparently been lost.

The idea of this book has a direct link with the so-called Xynthia event mentioned above. It was indeed suggested by the first author whose parents had been hit by the storm in the small town of Boyardville (Oléron Island) and experienced confusion following the event concerning the way in which expertise was used to delineate areas where houses at risk were to be destroyed and links with current policies were far from being well explained and understood by concerned citizens. The very same event triggered discussions at EU level regarding the consideration of storm surges in the context of the EU Flood Directive and Integrated Coastal Zone Management related measures. It also encouraged the EU to design research topics in the 7th Framework Programme on Research and Development to better understand the mechanisms of such disaster-prone events and to provide an improved knowledge base supporting existing and developing policies.

One of these projects was the so-called MICORE project (Morphological Impacts and COastal Risks induced by Extreme storm events) which aimed to develop a prototype Early Warning System (EWS) for predicting coastal storm risk. This paved the way for the currently running RISC-KIT project (Resilience-Increasing Strategies for Coasts – Toolkit) which aims to develop methods, tools and management approaches with links to EU and national policies, and which takes the historical dimension into consideration.

This book therefore starts by discussing the current regulatory framework related to coastal storms and flooding, with considerations about the need to develop a "culture of risks". It includes international and EU policies in the area of natural disaster reduction, civil protection and adaptation to climate change. Chapter 1 written by Philippe Quevauviller also discusses the need for better communication of scientific knowledge and policies, with an accent on interactions among different actors and implementation needs.

The second chapter focuses on the technical and scientific aspects related to the assessment of coastal storm risks. It is built on Paolo Ciavola's experience in the coordination of the above-mentioned MICORE project and his involvement in the RISC-KIT project, leading the construction of storm impact databases in cooperation with Emmanuel Garnier, the author to the third chapter. The author provides a synthesis of gathered knowledge on the evaluation of coastal vulnerability, including mapping and modeling, and a summary of recommended disaster risk reduction measures.

Finally, the historical knowledge dimension is fully developed in Chapter 3 written by Emmanuel Garnier. It focuses on understanding the mechanisms which may explain the magnitude of the Xynthia disaster occurring in a developed country and important member of the European Union. In particular the chapter focused on the example of the coastal town of La Faute-sur-Mer where 29 people of the same district drowned in the night. For this purpose, the chapter discusses the trajectory of vulnerability followed from the 18th century until 2010, with the objective of showing how a lesson learned, extracted from historical documentation, could have reduced this vulnerability considerably, like it has already done in certain countries of Northern Europe.

The book is aimed to address a wide readership covering the policy-making community, scientists and academia, practitioners as well as regional and coastal city authorities. It is hoped that the language will also be accessible to provide background information to citizens.

Philippe QUEVAUVILLER
Paolo CIAVOLA
Emmanuel GARNIER
January 2017

1

Coastal Storms and Flooding: Regulatory Framework and Science–Policy Interactions

1.1. Introduction

Coastal storms and flooding are areas of increasing concern, owing to growing urbanization and the related increased vulnerability to extreme climate events [IPC 14]. The increasing exposition of populations to such hazards leads to challenges linked to the development and implementation of regulations and management practices which take into account the adaptation to climate change. Actions related to improved risk prevention and reduction of climate-related hazards are embedded nowadays into international, European Union and national regulations and related management frameworks. In the case of coastal zones, however, although the prevention of hydrometeorological hazards is recognized as a priority at national level in many countries, the precise definition of these risks and hence their relations with the existing policy framework remains unclear [LAR 15]. The identification of coastal hazards (mainly storms and floods) concerns several

---

Considerations expressed in this book represent the sole views of the authors and do not reflect the formal position of any EU institution

domains other than the environment; in particular, urbanism, forestry, the insurance sector, tourism and civil protection, which are addressed by different types of regulations and codes of practice. This fragmentation is reflected from the EU to the national/regional levels. In addition, coastal risks are not subject to specific regulations despite being often mentioned, e.g. in research projects (see section 1.5) and texts related to territorial management plans.

Natural hazards in coastal zones can be separated into different types of risks [CAN 14], either of marine origin, e.g. submersion/flooding, storms, erosion, tsunamis, waves due to tornados, or of geological nature, e.g. land subsidence, earthquakes, landslides. In the context of this book, the focus will essentially be on coastal storms and flooding related to marine submersions, which may result from a combination of different phenomena such as wind, waves, high tide coefficients, low atmospheric pressures (leading to an elevation of the marine level) and abundant rainfalls. Marine submersions are considered as cumulative phenomena leading to either (or both) rapid or progressive flooding. The case of the Xynthia storm, which occurred along the West coast of France in February 2010, deals with both cases, i.e. a flash flood in the North (La Faute-sur-Mer) and progressive flooding in the South (Oleron island, Fouras, etc.). In each case, such submersion may lead to displacement of populations, significant modifications of land occupations and uses, losses of land as well as various environmental perturbations in wetlands and lagoons with modifications of the local ecosystems (e.g. salinization of freshwater resources and disappearance of beaches). Socio-economic and environmental impacts related to natural hazards in coastal zones, and notably coastal flooding, are therefore highly significant and subject to many controversial debates and studies. In reference to Xynthia, consequences of this storm generated many enquiries, scientific studies, historical research, etc., which all pointed out limitations of knowledge about coastal risks and their social nature. Adversely, policymakers and politicians did not express themselves very much at all, despite media controversies flourishing in relation to inadequacies of the regulatory framework. Chapter 3 refers to this specific case which is also described in details (in French) in the literature [GAR 13, LAR 15]. The present chapter provides a general background about the existing regulatory and EU-funded scientific framework and discusses difficulties in establishing interactions among the scientific and policy-making communities.

According to IPCC [BAT 08], observational records and climate projections provide abundant evidence that observed warming over several decades has been linked to changes in the large-scale hydrological cycle (e.g. effects on atmospheric water vapor content and changes of precipitation patterns with consequences on extreme floods and droughts). The consequences of climate change, in particular the increased frequency and severity of extreme hydrometeorological events, may alter the reliability of current water management systems. While quantitative projections of changes in precipitation, river flows and water levels at the river-basin scale remain uncertain, it is very likely that hydro(meteo)logical characteristics will change in the future. These considerations lead to the development of a complex policy framework which plans adaptation and mitigation options to tackle impacts of global warming on water resources and risks to society and assets. These options are closely linked to a range of policies. This chapter gives an outline of some policies relevant to hydrometeorological events, with no pretention of exhaustiveness. Some considerations have been adapted from previous publications [QUE 11a, QUE 11b, QUE 14].

1.2. Natural hazards and risks in coastal zones: needs to build-up a "culture of risks"

1.2.1. *Introduction*

Coastal zones were unoccupied for centuries as they were considered as dangerous areas by populations. This has changed only recently with the establishment of populations along coastlines, increasing their exposure to natural hazards (e.g. flash floods, seismic risks associated with tsunamis, marine submersions and erosion). Nowadays the littoral zone is very much in-demand and is confronted with a growing demography. The United Nations even estimates that, at the 2020 horizon, around 80% of the world population will live in a territorial stretch less than 100 km from the sea. As a result, natural hazards in coastal areas will increasingly threaten human societies, their assets and activities. These threats will be exacerbated by climate change which has an influence not only the sea-level rise but also on rainfall regimes and storm severity [LAR 15]. In less than 20 years, the awareness about climate-related risks has led to a huge development of research activities (see section 1.5).

A natural risk is generated by the conjunction of a threatening phenomenon which is identified as a hazard, and related human challenges characterized by their vulnerability to this hazard. In a way, human societies' awareness of natural hazards has certainly played a major role in the development of environmental policies, originally aimed toward the prevention of disasters rather than the protection of the environment itself. This policy sector has hence led to establishing founding principles of what may be referred to as a "culture of risks", which is being developed at policy level as well as in the collective consciousness [LAR 15].

In parallel, coastal zones present the paticularity of gathering several environmental compartments, which generates legal discussions that are reflected in specific environmental policies, with the risk of fragmentation. Considering the diversity of natural hazards, their intensity and potential impacts which may vary from region to region, the regulatory framework and its implementation may have different socio-economic and environmental dimensions, and the specific regulations have to be properly coordinated in order to manage the risks efficiently.

1.2.2. *Contribution of environmental policy to the development of a "culture of risks"*

According to Michelot in [LAR 15], the "culture of risk" can be defined as "the ensemble of perceptions and behaviours adopted by a society facing risks". It is associated with the society's memory about risks, i.e. a system of maintenance of the legacy of knowledge from the past and its consideration for reduction of today's vulnerability. This presents the idea of highlighting the need to better understand hazards and their impacts, and of developing the information, the education and the memory of risks. In a sense, developing a "culture of risks" is closely related to a democratic decision-making process on both an individual and collective scale. Every citizen should have the possibility to freely access information about the risks to which he/she is exposed as well as objective information on what is done to protect his/her security. The Civil Society, enterprises and representatives of national authorities should also be in the position to respond to citizen's expectations regarding their security.

This awareness carries greater significant nowadays with the increased frequency of natural hazards, some of them leading to disasters. This risk awareness is, however, not new. In the marine sector, the risks for navigation were well identified and related to dangers that were not linked to human actions [EWA 86]. Until the 19th Century, the notion of risk was linked to natural events. This changed with the industrial revolution, which led us to consider human activities as generating risks as well which had to be understood, managed and even anticipated [LAR 15].

Before speaking about environmental policies and their links to the "culture of risks" for coastal areas, it is useful to recall the role of policies in the establishment of this culture of risks. In a sense, we might argue that policies should include risk components and related objectives. This is the principle followed more largely by the DPSIR approach (Driver-Pressure-State-Impact-Response), which includes the notion of risks and responsibilities regarding the "responses" to address them. Environmental policies have the peculiarity of repositioning the relationship with risks in a holistic manner. In other words, many different branches of policies are mobilized in a given area to identify, assess and more generally manage risks or the related responsibilities. An example of integrated policy is the EU Water Framework Directive, which follows the above mentioned DPSIR principle, starting from risk assessment and mapping, to monitoring responsive actions to meet well-defined objectives [QUE 11a]. An important aspect which is not easy to tackle is the way society handles risks in the light of a policy framework, which is often far from people's direct prerogatives. This embeds not only risks for the environment, but concerns policies related to working rights, consumer rights and health as well which all have to be taken into consideration regarding risk management, in particular prevention.

Environmental policies may be a useful way to better frame the question of risk. Indeed, the above-mentioned DPSIR approach readily addresses risks and their impacts, taking societal (and economic) impacts into account at different scales (from international, national to local). When reading international, European and national environmental policies in general, it becomes clear that risks are being addressed with a view to limit their impacts on the environment. The terminology may, however, differ in relation to different situations, e.g. "potential impacts", "major accidents", "disasters", etc., which often leads, notably at EU level, to a hierarchy of actions related to responses to identified risks.

Policies related to natural disasters have been developed in a fragmented way, concerning soil, land use management, environmental and civil protection policies. This fragmentation illustrates the difficulty to build up a "culture of risks" through the development and implementation of policies which are themselves sectoral and elaborated in different contexts. At European Union level, the first Environment Action Programme in 1973 included "problems to which coastal zones are confronted" in relation to urbanism and land use management. Later, the EU used the terms of threats or disasters affecting coastal areas, and the term "risk" become associated in particular to flood management. The Green Paper released in 2006 [EUR 06] developed for the first time a systematic approach for identifying coastal risks, either of natural or accidental nature. In the international policy context, the term "risk" mainly appears in the framework of cooperation in case of critical situations for fighting against marine pollution, implying that a capacity exists to evaluate the extent of the pollution risk and its impacts.

Environmental policies rely on a series of concepts and principles which are driven by new expectations of the society at risk. In this context, precaution and prevention principles are built up in relation to risks. Regarding prevention, policies have been developed in many different branches, including social policies, which concern well-identified risks. The precautionary principle, however, applies in cases of potential risks of which the occurrence is hypothetical [LAR 15]. From these two principles, various policies have been developed at international level, including EU regulations and involve initiatives related to knowledge development, transmission and access to information. In the international policy framework (e.g. Sendai Framework for Action described below), international cooperation is closely linked to actions related to prevention and precaution.

These principles have extended to environmental policies, but also to other fields such as health policies. Prevention and precautionary principles are complementary with the principle of information and participation of the public in environmental decisions. In this respect, environmental policies are closely related to the Aarhus Convention (1998) on information access, public participation in decision-making processes and access to justice, principles which have been embedded into environmental policies such as the Water Framework Directive. This convention which can be considered historical in environmental international policy requests Member States to gather information and make it accessible to the public. Public participation is guaranteed for all activities, which may have an important impact on the environment. Finally, the Convention foresees access to justice, notably for rejected information requests or contestations related to the legitimacy of any activity or legal act.

Other important principles provide a capacity for environmental policies to participate in the building up of a “culture of risks”. The principle of integration of environmental policies in public policies enables this culture of risk to be disseminated and made operational [LAR 15]. The question is now posed whether the environmental policy framework is well adapted to the protection of coastal areas.

Coastal zones represent more than 2 million kilometers of coastline for all the continents combined, including islands. The majority of these coastal environments are still in a natural state and are moderately or not at all affected by anthropogenic activities. In the areas which are strongly impacted by anthropogenic pressures, ecological challenges are very significant, in particular regarding biodiversity. In addition, owing to the predicted sea-level rise linked to climate change, human pressures will likely increase with additional impacts on natural habitats. Action programmes related to environmental policies have therefore been developed with the objective of protecting vulnerable coastal areas, including the conception of integrated management approaches which align littoral land use policies with the management of the marine environment. One of the commonly accepted definitions of the Integrated Coastal Zone Management (ICZM) relates to a governance system, which consists of a legal and institutional framework stating that land use development plans of coastal areas take due consideration of marine environment protection objectives and are developed with the support and consultation of stakeholders and populations concerned [EUR 09].

1.2.3. *Toward an integrated management of coastal zones*

The UN Convention on the Law of the Sea (UNCLOS) of 1982 [UNI 82] introduced the notion of marine environment with related principles of preservation and protection. It also set the principle of international and regional cooperation regarding prevention. Ten years later, the UN Conference on Environment and Development (UNCED) held in Rio de Janeiro [UNI 92] introduced, in Agenda 21, a chapter about the protection of oceans and all seas and coastal zones, and the protection and sustainable use of their biological resources. The integrated management of coastal zones has a prominent place in this agenda with the objective to generate new strategies of management and exploitation of coastal areas, from a national to a worldwide level, which are targeted toward precaution and prevision. One of the seven fields of intervention foreseen in this agenda concerned the assessment of "fundamental uncertainties concerning the management of the marine environment and of climatic changes" [UNI 92]. Conventions derived from this Summit also integrated the need to develop an ICZM approach for the management of risks, with a focus on the protection of biodiversity in line with the Ramsar Convention on the protection of wetlands (http://www.ramsar.org/).

1.2.4. *Regional instruments*

The first specific consideration of coastal zones at European level was made by the Council of Europe, which adopted Resolution 29 of 26 October 1973, setting the principles of sustainable management of coastal zones which aimed to be implemented at national level. The OCDE also looked into the ICZM question with a recommendation expressed by its Council on 23 July 1992. A wide range of regional organizations then became aware about the need to develop management methods and tools for coastal zone management with a focus on the coast–land interface. The European Union thus developed an EU Strategy for ICZM through a Communication adopted on 6 May 1994 [EUR 94] and a reflection paper 5 years later [EUR 99]. In addition, from 1996 to 1999, the Commission undertook a demonstration programme for ICZM which implemented a number of experimental practices through the LIFE project [LIF 96]. Two documents were adopted on the basis of the conclusions of this programme, first the Commission Communication and European Parliament on "ICZM: A strategy for Europe" [EUR 02] and second a recommendation by the Council and European Parliament about the implementation of ICZM in Europe (30 May 2002). In this recommendation, Member States were invited to develop a strategic approach to the consideration of the threat that climate changes represent for coastal zones, the dangers of sea-level rise and the increase of frequency and severity of coastal storms [LAR 15].

In 2006, the Green Paper on the future European Maritime Policy clarified the input of ICZM to the management of natural hazards. The paper highlighted that the links which exist among coastal and maritime issues at the land/sea interface represent a challenge at EU level for the success of the ICZM, which *ipso facto* turned it into a reference framework for risk prevention management. In particular, it dealt with the integration of risks in decision-making related to planning and investing. As a follow-up, a Commission Communication on ICZM evaluation in Europe (7 June 2007) underlined the increasing pressures on coastal zones and the risks linked to climate change, and questioned the purely environmental approach of ICZM. The communication stressed that ICZM should better take account of the objectives of sustainable development and social considerations. Finally, the EU Regulation 1255/2011 of the European Parliament and the Council of 30 November 2011, which established a programme in support of the development of an integrated maritime policy, confirmed that the maritime scope of ICZM embeds important instruments for the sustainable development of marine and coastal zones and contributes to objectives of ecosystem management and the development of land/sea relationships. From the strict viewpoint of natural hazards, the regulation encourages actions aiming to attenuate climate changes on the marine, coastal and island environment and underlines that particular attention will be given to the most vulnerable zones.

These vague dispositions are in contrast with specific actions set by the EU Directive relative to Flood Risks adopted in 2007 [EUR 07], referred to in section 1.3.2.4. The European Union is also involved in the Regional Seas

Conventions promoted by the Marine Strategy Directive 2008/56/EC of 17 June 2008 [EUR 08] which preceded the implementation of the Barcelona Convention for the protection of the marine environment and coastal zone of the Mediterranean, as well as the Oslo and Paris Conventions for the protection of the North-East Atlantic and the Helsinki Convention for the protection of the Baltic Sea (see [QUE 11b]) for further details about these conventions). It also recommends that Member States of a same marine basin collaborate and act together to respect obligations and commitments resulting from international agreements. Among these instruments, the Madrid Protocol on ICZM includes, in Article 5e, objectives to prevent and/or reduce the impact of natural hazards, in particular related to climate changes, which may be related to natural or human activities [EUR 09]. EU Member States obviously have to implement EU regulations that they have themselves adopted, along with international conventions. Some of them, however, transpose the minimum technical requirements of European Directives, which is the case, for example, for the EU Flood Directive in the case of, e.g. France [LAR 15].

In consideration of the above legal framework, it is worth mentioning that these instruments have often been developed on an *ad-hoc* basis without global perspective and means adapted to implementation at national level. As a consequence, the management of natural hazards in coastal zones relies on different evaluation steps and follow-up, which are not always foreseen in environmental regulations that essentially focus on environment protection. It should be noted that the increased awareness about ecological risks has led to a repositioning of the environmental legal framework for the management of coastal zones. It is actually questioned whether this legal framework disposes of instruments, concepts and means to face challenges related to a “culture of risks” [LAR 15].

1.2.5. *Emerging "culture of risk" beyond the environment legal framework?*

Risks affecting coastal zones present peculiarities despite sharing common features with fluvial risks, which are linked to hydrological phenomena. Maritime natural hazards include hydrometeorological, climatic and geophysical risks. Hydrological phenomena are primarily linked to the variations of sea level, for which we may distinguish several elements: the mean level, the level of theoretical tide and the height above or below the mean sea level (which corresponds to the differences between the theoretical tide and the observed water level). Tsunamis lead to long period wavelengths provoked by non-meteorological phenomena (e.g. submarine landslides, earthquakes and volcanic eruptions). Similarly, coastal risks, as they are considered in coastal risk prevention plans, are not really linked to hydrological phenomena. They include principally submersion risks, risks of coastline retreat, as well as risks related to invasion of dunes. The prevention of these risks may actually lead to a fragmentation of legal instruments. At the moment of the Xynthia occurrence particularly, numerous public and private actors considered that the risk of marine submersion was not comparable to any other risk of natural origin. These phenomena can actually be linked, e.g. erosion can lead to a reduction of natural coastal defenses, thus enhancing the coastal vulnerability to marine submersions [LAR 15].

Our knowledge about coastal risks is improving and guiding us toward studies of vulnerability of coastal zones, to not only build up knowledge and a memory of risk, but also an "intelligence of risks". With the support of research, we could develop conceptual approaches that would go beyond the possible "predictions" of threatening events by science, i.e. turning this conception into "social constructions" with the mobilization of instruments that would give more decision-making power to the people and enhance their responsibilities. Research trends related to these efforts are described in section 1.5.

Environmental policies have been developed in consideration of a conception of risks being as realistic as possible. Numerous legal instruments related to coastal risks underlined the importance of research and expertise as well as citizens' access to information. Studies carried out by IPCC in the field of climate change and the progress which has been and is currently being made thanks to their conclusions illustrate this approach. However, environmental policies have progressively developed into a precautionary approach which has introduced an element of uncertainty in the decision-making process. In [LAR 15], Michelot underlines that the way of perceiving and taking account of risks has been modified, moving toward a socialization of risks that has to be explored to face the ecological, social and economic consequences of disasters. The question is whether the perspectives opened by the environmental policy framework are adapted to the social challenges of modern society.

1.2.6. *Ecosystem services: a new orientation of environmental policies or an opportunity of risk socialization?*

The European Union has adopted an action programme, the objective of which is to stop the degradation of ecosystem services on its territory from now until 2020 [EUR 12]. At international level, research activities are multiplying in support of international conventions, in particular concerning biodiversity and wetlands (Ramsar Convention, see above). The action plan of the Convention on biodiversity includes among its objectives the aim to guarantee the continuous function of ecosystem services.

Ecosystem services are increasingly present in the legal framework and introduce new ways to tackle natural hazards with different interpretations of their impacts. The understanding of coastal risks in the light of ecosystem services, i.e. looking at services provided by nature, may modify the perception of these risks in that it modifies what could (or should) be the knowledge-base needed to manage these risks. However, by introducing the term "services" in place of "functions", one also introduces the question "to whom, and for whom?" This may, in principle, require a legal approach to establish responsibilities related to ecological damages but it does find responses, in particular in the framework of the EU Directive on Environmental Responsibility (2004), in which ecological services are defined as "functions assured by a natural resource in benefit to another natural resource or the public" (quoted by Michelot in [LAR 15]). The question is how do we choose services to be preserved and in relation to whom? What "level" of service do we need to take into account (international, national and local)? It is obvious that the challenges and interests are not the same for risk managers and policymakers. Ecosystem services undoubtedly lead to the modification of risk perception, but do not necessarily tackle ecological or social inequities nor preserve the security of people and goods in the event of coastal hazards. In addition, these services do not provide a consensual approach for risk management by the different actors concerned [LAR 15].

1.2.7. *The international society facing (coastal) natural hazards: between protection of human rights and challenges of international security*

According to the IPCC, the vulnerability of risks lies on three main components: the exposition to hazards and impacts of climate change, the resilience of the environment to these impacts and the response capacity of society. It is, however, established that human actions regarding land-use management as well as the perception of risks may enhance or reduce the exposition to natural hazards.

Facing large-scale risks, the international society has deployed a range of tools and instruments to respond in an international cooperation framework which goes beyond the sole environmental dimension set at the UN Conference held in Rio in 1992 and the Johannesburg summit in 2002 where one of the objectives dealt with the reduction of the number and effects of natural- and human-related disasters. The United Nations have primarily decided to adopt a 10 year work programme (1990–1999) leading to the Yokohama Action Programme for the reduction of disaster risks, which has been continued by the adoption of the Hyogo Framework for Action (2005–2015) of which the main objective was to substantially reduce the loss of human lives and damages related to (natural) disasters. This is now pursued by the Sendai Framework for Action adopted in March 2015 for another period of 10 years (see section 1.3). Even if these international conventions do not lead to specific responses to coastal risks, they set a priority for UN Member States to reduce disaster risks from prevention/ preparedness to response/ recovery. This action framework aims to include risk reduction actions in national policies and reinforce the coordination of actions engaged at international and national levels.

In the UN framework, a range of organizations work for the development of a "disaster culture" for the international society. The UNDP (UN Development Programme) strives to ensure that disaster risks are taken into consideration in the development programmes of UN Member States and that they are engaged in a process of risk and vulnerability reduction for the future. The UN Environment Programme (UNEP) supports the dissemination of environmental information and rapid alert systems related to environmental risks on an international scale. Other organizations such as UNESCO and the WMO (World Meteorological Organisation) also contribute under their remit to feed this "disaster culture" through the development of studies and scientific activities (including alert systems) which enable us to anticipate risks and hence to save lives and limit material damages.

In parallel, climate change has become an international security challenge, considering that most of the humanitarian crises at international level are related to events linked to extreme climatic events. Debates are on-going concerning the implications of climate change on international security, and on a "society of risk" to face threats of destabilization related to the impact of climate change conflicts about resources, territory losses and border conflicts, threats to critical infrastructures, migration related to environmental threats, tensions linked to energetic provisions, pressures on worldwide governance, etc. In relation to coastal zones, the intensity and probability of risks are exacerbated. Sea-level rise and the increase in frequency of natural disasters represent a serious threat to numerous regions in the world.

1.3. Policy background

1.3.1. *International policies*

While climate change's impact on water resources and proactive management efforts are recognized worldwide, it is only recently that the concept of a "global policy" dealing with prevention/adaptation measures related to hydrometeorological events linked to climate change has been discussed. This need has been clearly expressed with regard to other factors related to water resources by the UN Economic Commission for Europe (UNECE) in a guidance document published in 2009 [UNI 09]. The key messages were along the same lines as recommendations expressed in the IPCC Technical Paper on water [BAT 08], in particular concerning the negative impact on nearly all UNECE countries linked to increased frequency and intensity of floods. UNECE recommended that any policy needs to consider climate change (and related extreme events) as one of the many pressures on water resources (others include population growth, migration, globalization, changing consumption patterns as well as agricultural and industrial developments). Effective adaptation in this respect was stressed as well as the necessity of a cross-sectoral approach, on a trans-boundary level, of in order to prevent possible conflicts between different sectors and consider trade-offs and synergies between adaptation and mitigation pressures. The guidance insisted on the fact that legislation should not present barriers for adaptation and should be flexible enough to accommodate continuing environmental and socio-economic changes. Besides national legislation, a number of international agreements include provisions that can support the development of adaptation strategies (see below), and trans-boundary cooperation and policy certainly might build upon this basis. Finally, the UNECE guidance stressed that uncertainty should not be a reason for inaction and highlighted that action, knowledge and experience sharing, and research on adaptation should be pursued simultaneously and in a flexible way.

As mentioned in section 1.2, a key policy trend in the last decade has been represented by the Hyogo Framework for Action 2005–2015 (HFA), which was adopted by 168 nations committed to substantially reducing the loss of life and livelihoods from natural disasters (including coastal storms and floods). The implementation of the HFA has been under the responsibility of the United Nations International Strategy for Disaster Reduction (UN-ISDR) which is the central point in the UN system for the coordination of disaster reduction and the establishment of synergy among the disaster-reduction activities of the UN and regional organizations, and activities in socio-economic and humanitarian fields. This framework is now continued by the Sendai Framework for Action which has set priorities for the 2015–2025 period, among which the promotion of a better understanding of disaster risk management through the building, sharing and development of knowledge and the strengthening of the policy–science interface at local, national, regional and global levels [UNI 15]. The Sendai Framework for Action sets out an ambitious set of priorities to position disaster risk reduction as a key element of sustainable development efforts, to define further steps to reduce existing and emerging risks and foster disaster resilience. These objectives are supported by IPCC recommendations expressed in the special report on extreme events (STREX Report).

1.3.2. EU policies

A wide range of sectors and policies cover issues related to the security, safety and resilience of society (therefore including impacts of natural hazards such as coastal storms and floods) in a direct or indirect way, either by providing legally binding frameworks of actions by EU Member States in the form of Directives, general frameworks in the form of Communications or technical specifications in the form of Decisions, for example. Figure 1.1 gives an illustration of the different "families" of EU General Directorates in charge of various policies as well as Intergovernmental Agencies.

Crisis Management policies follow an integrated approach for the management of natural and man-made hazards focusing on disaster risk reduction (prevention and preparedness) and disaster response. The policy is mainly represented by the EU Civil Protection Mechanism (UCPM) represented by DG ECHO [EUR 13a], and the operational dimension is coordinated by the Emergency Response Coordination Centre (ERCC). Disaster risk management is also addressed through the EU Internal Security Strategy [EUR 10] and the resulting European Agenda on Security adopted in April 2015 [EUR 15] represented by DG HOME and Consumer Health Protection policies [EUR 13b] represented by DG SANCO). In addition, climate-related disasters are covered by environmental and climate policies (DG ENV, in particular the Flood Directive [EUR 07] and DG CLIMA through the EU climate change adaptation strategy [EUR 13c]). Finally, intergovernmental agencies are also involved in security policies, e.g. the European External Action Service (EEAS) – which implements the EU Common Foreign and Security Policy – and Europol – which is the EU Law Enforcement Agency. Both agencies assist EU Member States. There are also links with the Council Decision 2014/415/EU on the arrangements for the implementation by the solidarity clause Union, which covers response, situational awareness as well as analysis and threat assessment at Union level.

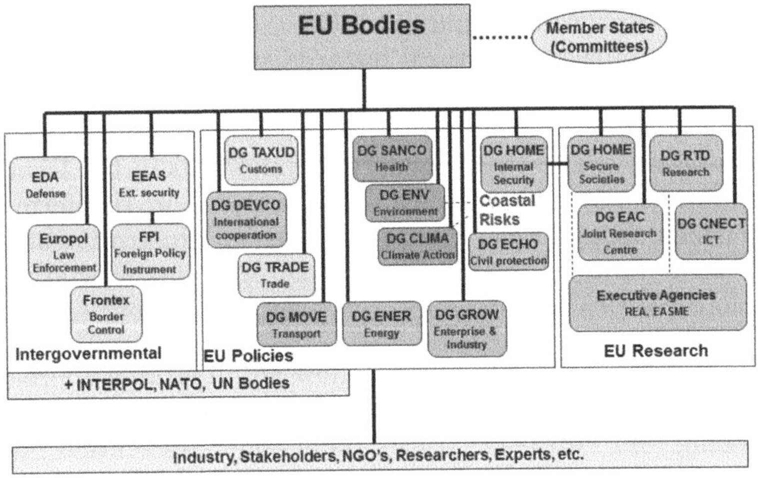

Figure 1.1. *The EU policy landscape. For a color version of this figure, see www.iste.co.uk/quevauviller/storms.zip*

Other key EU policies concern industrial competitiveness and innovation, namely, the EU Industrial Policy [EUR 12a] represented by DG GROW which aims to boost industrial competitiveness and innovation (thus the access to market of developed technologies) and the EU research policy represented by Horizon2020 (see section 1.5). To complement this figure, let us mention policies related to Energy Infrastructure and Transport Networks (DGs ENER and MOVE), Customs (DG TAXUD), Environment and Industrial Risks (DG ENV) and International Cooperation (DG DEVCO).

Complementary to EU policies, international policies are also active in Disaster Risk and Crisis Management. Their implementation represents a complex and ambitious challenge as they involve a wide variety of players, each Member State following specific national approaches (national action plans) for dealing with crises and organized differently in terms of disaster risk management capabilities. The EU framework represents a means and a real opportunity to discuss possible ways to improve coordination among the various national approaches and develop a common EU vision strengthened by a joint strategy in this field.

Compared to many international river basins worldwide with no legally enforceable management framework, the situation in the European Union is developing toward a robust, risk-based management system for tackling environmental hazards including coastal threats, with legal instruments being in place or development. This section examines concrete policy steps that are either implemented or being developed in Europe.

1.3.2.1. *EU Civil Protection Mechanism and related international policies*

The Union Civil Protection Mechanism (UCPM) under DG ECHO aims to facilitate reinforced cooperation between the EU and Member States and to facilitate coordination in the field of civil protection, in order to improve the effectiveness of systems for preventing, preparing for and responding to natural (including coastal storms and floods) and man-made disasters. It supports and complements the efforts of the Member States in the protection primarily of people but also of the environment and property, including cultural heritage, in the event of natural and man-made disasters, acts of terrorism and technological, radiological or environmental accidents, including marine pollution. Built upon these policy instruments, the UCPM is about developing an integrated approach to disaster management. The EU action is based on the principles of solidarity. The overall mechanism takes due consideration of laws and international commitments, and exploits synergies with relevant Union initiatives such as the European Earth Observation Programmes (Copernicus), the European Programme for Critical Infrastructure Protection (EPCIP) and the Common Information Sharing Environment (CISE). The mechanism is based around the Emergency Response Coordination Centre (ERCC) and the European Emergency Response Capacity (EERC) in the form of a voluntary pool of precommitted capacities from the Member States, trained experts, a Common Emergency Communication and Information System (CECIS) managed by the Commission and contact points in the MS. It also recognizes the role of regional and local authorities in disaster management. Outside the Union, disaster response is coordinated with the United Nations (in close interaction with the Sendai Framework for Action) and other relevant international actors related to humanitarian aid. Finally, the use of military means under civilian leadership as a last resort may constitute an important contribution to disaster response.

On a technical level, the UCPM is working toward a general policy framework on disaster risk prevention aimed at achieving a higher level of protection and resilience against disasters by preventing or reducing their effects and by fostering a culture of prevention. From this perspective, it promotes the review of risk assessment, risk management planning conducted at national/regional level and the development of an integrated approach, linking risk prevention, preparedness and response actions. The UCPM is also financing actions related to preventing, preparing for and responding to disasters, including civil protection training programmes, large-scale exercises, interaction of experts, prevention and preparedness projects (through annual calls for applications), logistical and transport support for response missions, deployment of coordination, etc.

1.3.2.2. *Critical Infrastructure Protection*

The new approach to the European Programme for Critical Infrastructure Protection (EPCIP) under DG HOME aims to ensure a high degree of protection for EU infrastructures and increase their resilience against all threats and hazards [EUR 13d], including coastal threats. The sector-focused approach of the programme represents a challenge to a number of EU Member States as in practice the analysis of criticalities is not confined to sectoral boundaries and follows rather a “system” or “service” approach (e.g. hospitals). There is a need for the development of a cross-sectoral approach. In practical terms, development of preparedness strategies is based around contingency planning, stress tests, raising awareness, training, joint courses, exercises and staff exchange. The programme also promotes dialogue between the operators of the critical infrastructures and those who rely upon them in order to better prepare responses to events affecting European critical infrastructures. The gaps identified in the review of the EPCIP led the Commission to incorporate its new approach in the implementation of the EPCIP in 2013, with a greater focus on interdependencies and proposing practical work with four critical infrastructures of a European dimension (Eurocontrol, Galileo, the electricity transmission grid and the gas transmission network). These are linked to the trans-European energy infrastructure as such and the EPCIP stipulates that the Union's energy infrastructure should be upgraded in order to prevent technical failure and to increase its resilience against such failure, natural or man-made disasters, adverse effects of climate change and threats (including coastal storms and floods, and their possible cascading effects) to its security. It is also related to EU transport policies covering a wide range of security and safety policies in the air, road, maritime and rail areas which all relate to technical standards for preventing/detection risks and responding to major threats, including natural disasters.

1.3.2.3. *EU Adaptation Strategy to Climate Change*

The EU Adaptation Strategy to Climate Change under DG CLIMA highlights the consequences of climate change and the need for adaptation measures. It focuses on early, planned and coordinated action rather than reactive adaptation. The communication highlights the need for systematic exchanges of best practice on how to best adapt to climate change. The strategy takes account of global climate change impacts such as disruptions to supply chains or impaired access to raw materials, energy and food supplies. The overall aim is to contribute to a more climate resilient Europe by enhancing the preparedness and capacity to respond to the impacts of climate change at local, regional, national and EU levels, developing a coherent approach and improving coordination. This strategy is closely linked to national adaptation strategies, which are considered as recommended instruments by the UN Framework Convention on Climate Change (UNFCCC). Close coordination between climate change adaptation and disaster risk management/policies is also required. Development is planned based on guidelines of minimum standards of good practice for disaster prevention.

1.3.2.4. *Water and Marine policies*

Linked to the above, specific policy instruments are in place in the water sector related to extreme hydrometeorological events such as floods and droughts under DG ENV. Complementing the Water Framework Directive (WFD), flood prevention and management are tackled by the Flood Directive which requires EU Member States to assess and manage flood risks, with the aim of reducing adverse consequences for human health, the environment, cultural heritage and economic activity associated with floods in Europe [EUR 07]. This directive has to be coordinated with the implementation of the WFD [EUR 00] from the second river basin management plan onward (2015–2021). It therefore provides a comprehensive mechanism for assessing and monitoring increased risks of flooding (including coastal floods), taking into account the possible impacts of climate change, and for developing appropriate adaptation approaches.

In line with the subsidiarity principle, this directive leaves a large margin of judgement and implementation to the EU Member States, providing a framework which sets the method to be followed while imposing a holistic approach to risk management. It initially involves the identification and reporting of flood risk areas followed by a preliminary evaluation of the risks and then the reduction of consequences of floods. Besides the texts oriented toward the management of coastal and/or marine areas, the EU has developed a network of protected areas through the Natura 2000 network, an application of the Bird Directive of 2 April 1979, which enables the classification of zones requiring special protection for territories most significant for the conservation of wild birds. The Habitat Directive of 21 May 1992 complements this framework with the creation of specific conservation zones, among which, coastal habitats.

Finally, while the protection of the (coastal) marine environment is covered by the WFD, EU environmental policymakers considered there was a lack of strategy underpinning the policies to protect the marine environment. A strategy was thus developed in the sixth Environmental Action Programme (2002–2012), which resulted in the establistment of environmental objectives for the marine environment. The related protection regime is regulated under the EU Marine Strategy Directive [EUR 08].

1.4. Science–policy interactions

1.4.1. *Scientific foundation of coastal risk policies*

The need to strengthen links among scientific outputs and policy-making activities in overall water-related risk management has been subject to on-going debates in the last 10 years, and specific discussions in the water and marine sectors have examined concrete developments [QUE 10], which can be extrapolated to coastal risks. They tend to show that a conceptual framework for a science–policy interface among scientists, policymakers and stakeholders is required *inter alia* in the water and marine sectors. This section explores general considerations about science and policy interfacing needs (following on from [QUE 07]). In a first instance, let us recall that key steps of the "environmental (including marine and coastal) policy chain" related to protection against pollution and natural hazards are based upon a scientific foundation and basic technical knowledge; these steps can be summarized as follows:

– Describe what you want to protect;

– Measure or describe actual (pre-event) situation;

– Define level of protection according to well-defined objectives (related to infrastructures, assets and populations);

– Identify pressures;

– Quantify relationship between pressure and environmental response;

– Quantify relationship between social and economic cost and pressure;

– Identify lowest cost pathway;

– Define policy instrument;

– Implement the policy instrument and assess response;

– Take appropriate measures (control and remediation);

– Review policy on the basis of scientific/technological progress.

The reliability of the overall chain will indeed depend upon the effectiveness of the integration of scientific and technological knowledge in a timely manner at each step of the policy development, implementation and review. The knowledge of "environmental interfaces", i.e. interactions between different compartments, represents a basic feature for understanding the impacts of natural or anthropogenic pressures on various (marine) environmental compartments. It therefore has a direct impact on the way policy and related monitoring are designed, developed and implemented. This knowledge should, in principle, be tackled in a "holistic" manner. In other words, it is hardly possible to understand the overall impact of a specific pressure on coastal zones by looking only at one compartment [QUE 07]. Coastal risks depend upon a variety of environmental factors such as the climate, hydrology (water flows), hydromorphology, etc. In this respect, it is hardly possible to understand a given risk by looking at a single environmental compartment and through only one discipline. To date, knowledge of environmental interfaces is still limited by the lack of sufficient multidisciplinary studies. Progress is on-going in the framework of various EU-funded projects (see section 1.5), but the scientific foundation is not considered to be sufficiently developed to be able to effectively assess the effectiveness of coastal risk-related policies in a holistic context. It is worth mentioning that, among research projects and related on-going activities, gathering of an increasing number of monitoring data (linked to EU policies and/or international programmes) and the development of models now provide a much better vision of the problems to be tackled and the way to approach them.

1.4.2. *EU Scientific framework in support of coastal risk-related policies*

The Treaty establishing the European Union indicates that Research Framework Programmes have to serve two main strategic objectives. First, they provide a scientific and technological basis for industry and encourages its international competitiveness. And second, they promote research activities in support of other EU policies. To this end, Framework Programmes are designed to help solve problems and respond to major socio-economic challenges faced by society. The Research Framework Programme (FP) is the European Union's main instrument for funding research and development. In this context, the European Commission has been supporting marine (and coastal) research for several years through its successive Framework Programmes (FP) for Research and Technological Development (RTD). The FP aims to foster scientific excellence, competitiveness and innovation through the promotion of better co-operation and coordination. It also aims to produce advances in knowledge and understanding, and to support the implementation of related European policies. The FP is implemented through open 'calls for proposals', and successful projects are selected after an evaluation procedure carried out with the help of external independent experts.

Priority areas reflecting EU research needs are sectors such as health, food and agriculture, information and communication technologies, nanosciences, energy, transport, socio-economic sciences, space and security. Environment and climate change is also part of these priorities. More specifically, the research areas address pressures on environment and climate, impacts and feedback, environment and health, conservation and sustainable management of natural resources, evolution of marine environments, environmental technologies, understanding and prevention of natural hazards, forecasting methods and assessment tools, and earth observation. Details on the current framework programme (Horizon 2020) and examples of EU-funded projects are given in section 1.5.

1.4.3. *Identification of research needs in the coastal risk-related policy sectors*

It is not always possible to establish a clear distinction between "basic" and "applied" research. Also the timing aspect (short-, medium- and long-term) is intimately linked to the way research instruments are operated. The identification of research need is of course fed by advances in scientific knowledge, but is also directly influenced by the evolution and requirements of policies. The need for ensuring coincidence of research and policy agendas may depend upon the stage of development of the policy in a given thematic area. In this respect, one may distinguish three different categories of needs in the policy sector related to coastal risks, depending on timing considerations [QUE 07]:

– *Short term* (~1–2 years): needs basically concern the accessibility of research knowledge required for the development of policies on a short-term basis. Timing is not adapted to develop new types of research (unless very specific needs are identified, which may be developed in a 6–12 months period). Policy development also requires efficient and user-friendly access to background scientific information and archives; a typical example is thematic strategies such as the ones covered by the EU Environment Action Programmes. In this context, the time needed for the design, approval and operation of *ad hoc* calls for proposals makes it difficult to respond to short-term research needs, e.g. related to a coastal storm or flood requiring scientific input. In other words, a specific research need expressed at a given time will rarely be met through a project selected under a call for proposals the year after. Therefore, to date, such needs may only be tackled through research actions carried out by, e.g., national research organizations, having been identified in their annual work-programme. In the future, short-term needs could also be partially fulfilled through a coordination of national research calls for proposals coordinated by Joint Programming Initiatives.

– *Medium term* (~2–5 years): the timing of medium-term research is adapted to responses of needs expressed in relation to the implementation agenda of well-defined policies (representing a "stable platform" for building strong partnerships among policy implementers, the scientific community and various stakeholders). This is the case of the framework directives (e.g. the WFD, the Marine Strategy Directive, etc.) in support of which research activities have been carried out since the time of their adoption, in response to needs linked to, e.g., analysis of pressures and impacts. For the forthcoming policy milestones, the formulation of medium-term research needs will have to take into account technical requirements related to the implementation of the policies and their revision. This requires that a detailed description of research needs is made by policymakers and that a follow-up of projects is done in close coordination with the scientific community to guarantee a successful uptake and application of research to the policy-making process.

RTD projects running over a 2–3 year period may also fulfill medium-term research needs.

– *Long term* (~5–10 years): scientific progress in this respect supports either policy milestones which are clearly identified at the 10 year horizon, or the review process of the legislation. Long-term research needs may be linked to the development of action programmes. They may also concern the review process of the technical requirements detailed in a given directive, e.g. in the case of the WFD or the EU Flood Directive, the end of each River Basin Management Plan cycle (6 years). Research activities may respond to either well-defined milestones of the thematic policies or review of the legislation.

1.4.4. *Interactions with the scientific community*

At the start of projects which have been identified as relevant to coastal risk-related policies, there is certainly a need to clarify policy issues by describing the aims, milestones and technical challenges to the RTD project coordinators so that they understand what the policy expectations are over the duration of their project. These exchanges of information/knowledge rarely occur, which may lead to divergent directions being taken by the projects in comparison to policy orientations [QUE 07].

1.4.4.1. *Synthesis needs*

At the end of research projects, the most critical issue is the way the scientific information is "digested" so that it may be efficiently diffused to policy end-users and possibly applied. This integration phase is certainly the weakest link of the science–policy chain. Indeed, only a small percentage of RTD projects are known by policy implementers, which illustrates the need to not only improve awareness about RTD outputs, but also to encourage policy actors to reflect on research needs linked to their portfolios. This may be translated into needs to carry out synthesis works in the form of "policy digests" (addressed to the scientific community from the policy implementer's side) and "science digests" (prepared by the scientific community for the policy implementers).

1.4.4.2. *Exchange platforms*

As a follow-up to research or capacity building projects, useful interactions may occur at yearly meetings. Participation of policy officers in all project meetings may not be practicable due to a lack of resources, but efforts are needed to organize regular joint meetings focusing on specific themes. This is already happening in the WFD sector [QUE 07] and should be systematized.

1.4.4.3. *Toward a "science–policy interface"*

At the present stage, efforts are lacking for presenting results of research and demonstration projects in a form that policymakers may easily use, e.g. "science-digest" policy briefs. On the reverse side, the consideration of research results by the policy-making community is not straightforward, mainly for political reasons and difficulties integrating the latest research developments in legislation. The difficulty is enhanced by the fact that the policy-making community is probably not defining its role as "client" sufficiently well. In other words, the dialogue and communication are far from being what one would hope to ensure an efficient flow of information. In this respect, improvements could be achieved through the development of a "science–policy interface" based on a coordination of relevant programmes/projects with direct relevance to the policy implementation. In other words, strategies should identify needs for short-, medium- and long-term scientific developments and should establish an interface so that R&D results are synthesized in a way that can efficiently feed the implementation and further review of policies.

This interface should include:

– a screening phase evaluating which type of research is needed (background information or tailor-made research and demonstration) in accordance with the policy step of concern (e.g. implementation issues and reviewing). This is already happening through regular contact between Commission services and the scientific community;

– a mechanism to ensure that the most promising research projects in support of the policies are "validated" through demonstration activities, disseminated efficiently and applied at the appropriate level (regional, national or EU);

– a management scheme involving both scientists and policymakers to discuss the corresponding research and policy agendas from the very beginning in order to ensure a more structured communication at all appropriate levels of policy formulation, development, implementation and review. This is hardly operational to date.

More than dissemination and application, the interface should establish strong links between the different funding mechanisms existing at EU level and the thematic policies. This would enable us to promote pilot projects combining the implementation of results of successfully completed research or demonstration projects with the implementation of related policies. This would allow us to form new and innovative partnerships by combining various instruments (research, capacity-building, structural funds, education and training) and regional/national funding mechanisms, as well as the establishment of a collaborative partnership involving scientists, policymakers, managers and other stakeholders, for the effective integration of science outputs into policy and management decisions. At present, however, examples show that such coordination is still not operational. A developing "Community of Users" (see section 1.6) might facilitate an improved interfacing in the future.

1.4.5. *Science-based development of an integrated coastal risks policy*

One may ask the basic question: is our scientific (multi-disciplinary) knowledge sufficient to develop a more integrated policy able to efficiently tackle coastal risks? The on-going discussions show that the scientific base is likely still not sufficiently consolidated at this stage, but that a tight coordination mechanism and tailor-made developments in the H2020 programme (see section 1.5) could lead to the establishment of an operational science–policy interfacing mechanism at the 2020 horizon.

Apprehension about scientific uncertainty is growing in importance. Such misgivings seek to invoke the standard of evidence that 'guilt' must be demonstrated 'beyond a reasonable doubt'. However, given the complexity of coastal threats and pathways, this would mean that the reality of coastal risks would not be accepted until they had actually happened. This is against the prevention principle and is not acceptable. In the light of the precautionary principle, however, where there are threats of serious or irreversible environmental damage, lack of full scientific certainty should not be used as a reason for postponing measures aimed at preventing environmental degradation.

In the context of coastal risks, the following observations can be made:

– Coordinated reporting and data sharing about coastal risks should constitute the core basis for policy implementation and review within the next decade. In the light of increasing considerations about climate change (and its impact on coastal risk management), one may wonder whether this data sharing should not expanded at a global scale in close interactions with the Copernicus programme;

– Building-up coastal databases should enable us to test/validate existing models and develop new models able to better evaluate coastal risks linked to climate extreme events, and thus better evaluate the efficiency of policy responses. This is closely linked to the knowledge-based considerations expressed in this chapter;

– Risk assessments and programmes of measures need to be coordinated in the light of effective implementation of policies in force (in the different sectors).

The consequence of better integration of scientific knowledge and policies should have a positive impact on the way coastal risk (disaster) management will be implemented in future, i.e. a better knowledge-based appraisal of risks in the context of concerted planning (e.g. at river basin level) should facilitate the design of surveillance and detection programmes (avoiding duplication and focusing on specific features) and the elaboration of proper prevention/ preparedness strategies as well as response programmes. The way international policies (such as the Sendai Framework for Action) are being developed pave the way for an increasing integration, which should be pursued and linked to a sound and validated scientific foundation.

1.5. Research trends

1.5.1. *Introduction*

Research related to coastal threats (storms and floods) is needed to improve understanding and modeling of climate changes related to extreme hydrometeorological events on scales that are relevant to decision-making (possibly linked to policy). At present, scientific information about coastal risks and their links to climate change is not sufficient, especially with respect to understanding and the assessing of key drivers and their interactions in order to better manage and mitigate risks and uncertainties. Arising questions concern scientific outcomes that are sufficiently mature to be taken aboard policy development and which are key research topics that need to be addressed at European level.

Scientifically sound data and other information are essential for making climate projections while reducing their uncertainties, particularly for vulnerable groups and regions, in relation to coastal storms and floods. This includes issues encompassing all aspects of the hydrometeorological cycle; taking into consideration the needs of end-users and including social and economic information. For instance, early warning systems are essential for preparedness for extreme weather events and should be developed at the trans-boundary level. They should also be closely linked to seasonal and long-term climate and weather forecast systems, as well as monitoring and observation systems.

1.5.2. *EU-funded instruments*

The European Commission is funding research through its Framework Programme for Research and Technological Development. In this context, projects of the sixth Framework Programme (2002–2006), in particular, projects funded under the "Global Change and Ecosystems" sub-priority and the seventh Framework Programme (2007–2013) or FP7, in particular, projects funded under the "Environment (including climate change)" theme, largely contributed to gathering knowledge relevant to natural hazards knowledge base, including for coastal storms and floods. As highlighted in section 1.3.2, EU research funding is orchestrated by different "research families", namely, various programmes of DG RTD, DG CNECT and DG HOME (see Figure 1.1), as well as research actions undertaken by the Joint Research Centre (JRC). Other funding instruments focus on capacity-building and training (e.g. prevention, preparedness and response projects in disaster risk management funded by DG ECHO and security-related projects funded by DG HOME), but they will not be discussed in this book. While research programming and policy responsibilities lay with the respective General-Directorates of the European Commission, the management of projects is increasingly delegated to "sister" agencies, namely, the Research Executive Agency (REA) and the Executive Agency for SMEs (EASME).

Horizon 2020 is the largest EU Research and Innovation programme ever with nearly €80 billion of funding available over 7 years (2014–2020) – in addition to the private investment that this money will attract. It promises more breakthroughs, discoveries and world-firsts by taking great ideas from laboratory to the market. It is the financial instrument implementing the Innovation Union, a Europe 2020 flagship initiative aimed at securing Europe's global competitiveness. By coupling research and innovation, Horizon 2020 is helping to achieve this with its emphasis on excellent science, industrial leadership and tackling societal challenges. The goal is to ensure that Europe produces world-class science, removes barriers to innovation and makes it easier for the public and private sectors to work together in delivering innovation. Horizon 2020 will contribute to the implementation of many policy goals, among which the Union Civil Protection Mechanism, environmental and climate policies. The primary aim of the Work Programme on "Secure societies – Protecting freedom and security of Europe and its citizens" is to enhance the awareness, preparedness and resilience of our society against natural and man-made disasters.

The current EU Framework Programme for Research and Innovation is built upon achievements of the seventh Framework Programme, whose mapping focused on and embedded several programmes of direct and indirect relevance to secure, safe and resilient societies, namely:

– health, demographic change and wellbeing;

– food security, sustainable agriculture and forestry, marine and maritime and inland water research, and the bioeconomy;

– secure, clean and efficient energy;

– smart, green and integrated transport;

– climate action, environment, resource efficiency and raw materials;

– Europe in a changing world – inclusive, innovative and reflective societies;

– secure societies – protecting freedom and security of Europe and its citizens.

1.5.3. *Examples of research trends*

Most of the research projects listed in this section directly or indirectly support policies related to coastal storms and floods in the areas of disaster prevention, preparedness and response.

1.5.3.1. *Climate change impacts on the water environment and cycle*

Specific research on climate change impacts on the global water cycle has been carried out under the FP6 WATCH (global change and water) project (www.eu-watch.org) which united different expertises (hydrologists, climatologists and water use experts) to examine the components of the current and future global water cycles, evaluate their uncertainties and clarify the overall vulnerability of global water resources related to the main societal and economic sectors. Current generation of large-scale models (e.g. climate models and global hydrological models) were tested against more detailed (hydrological) models to explore their ability to predict droughts and floods. Other FP7 projects on climate change impacts on water resources are described by Quevauviller [QUE 12].

1.5.3.2. *Floods research*

The project most relevant to flood research carried out within the years 2004–2009 at EU level in support of the Flood Directive is certainly the FP6 FLOODsite (integrated flood risk analysis and management methodologies) Integrated Project, which gathered interdisciplinary, integrating expertise from across the environmental and social sciences, as well as technology, spatial planning and management. The notion of "integrated" flood risk management evolved from flood defense to flood risks being managed, but not eliminated. The project developed robust methods of flood risk assessment and management and decision support systems which were largely tested in pilot sites. Regular contacts with EU policymakers enabled the policy community to be informed about progress on flood risk management. More than 100 research reports were made available for public download on the project website.

Albeit not focusing on coastal floods, the FP7 IMPRINTS project (http://www.imprints-fp7.eu) developed methods and tools to be used by emergency agencies for improving the preparedness and the operational risk management for flash floods and debris flow generating events, as well as contributing to sustainable development through reducing damages to the environment. Impacts of future changes, including climatic, land use and socio-economic were analyzed in order to provide guidelines for mitigation and adaptation measures. Systems were tested on five selected flash flood prone areas supervised by risk management authorities and utility company managers in duty of emergency management. One major result of the project will be an operational prototype including the tools and methodologies developed under the project. This prototype was designed under the premise of its ultimate commercialization and use worldwide.

Also in FP7, the FLOODIS project ("Integrating GMES Emergency Services with satellite navigation and communication for establishing a Flood information service" - http://www.floodis.eu/) provided accurate location-based application for portable devices, closing a critical gap for disaster management teams, civil protection and field/emergency response units to better address and mitigate crisis situations arising before, during and after heavy flooding. Access to open-source, location-based smart phone application for the general public to enable the capacity for individuals to take precautionary actions, therefore, vastly reduces the likelihood of human and economic loss. The project also considered rescuers relying on professional terminals and legacy communication channels. This combined Earth Observation and GNSS (GPS, Galileo, EGNOS/EDAS) technologies deliver alerts and interactive maps on flooding risk/events to users in the geographical area at risk.

1.5.3.3. *Risk assessment of climate-related hazards*

Complementing the above, risk prevention and reduction of climate-related disasters have been subject to major research efforts, namely, by the FP7 KULTURISK project (http://www.kulturisk.eu/) which developed a culture of risk prevention by means of comprehensive demonstration of benefits of prevention measures through enhanced memory and knowledge of past disasters, communication and understanding capacity of current and future hazards; awareness of risk and preparedness for future events. Measures included early warning systems, non-structural options (e.g. mapping and planning), risk transfer strategies (e.g. insurance policy) and structural initiatives. Focus was on water-related hazards with case on floods, debris flows and landslides, and storm surges.

More recently, the FP7 INTACT project ("Impact of Extreme Weather on Critical Infrastructures") assessed regionally differentiated risk in EU associated with extreme weather, identified, classified on EU wide basis critical infrastructures and assessed their resilience to extreme weather events. The project identified potential measures, technologies to consider, implement, be it for planning, designing and protecting critical infrastructures or for effectively preparing for crisis response and recovery. It brought together a community of climatologists, civil protection operators and meteorologists, with those of owners/operators of critical infrastructure planners to develop prevention of major disasters in cascading effects.

Within Horizon2020, several topics responded to research needs in support of the climate change adaptation strategy with consideration of extreme events, namely, topics dealing with "Science and innovation for adaptation to climate change: from assessing costs, risks and opportunities to demonstration of options and practices", "Mitigating the impacts of climate change and natural hazards on cultural heritage sites, structures and artifacts" and a study on "Impact of climate change in third countries on Europe's security", as well as "Crisis management topic 1: potential of current measures and technologies to respond to extreme weather and climate events". At the time of writing this book, research projects issued from these topics had not yet been selected and were aimed to start by mid-2016 for a period of 3–5 years.

1.5.3.4. *Coastal risks induced by storm events or flooding*

Coastal risks tend to be funded by several projects. The FP7 MICORE project ("Morphological Impacts and COastal Risks induced by Extreme Storm events" - https://www.micore.eu/) developed probabilistic mapping of the morphological impact of marine storms and produced early warning and information systems to support long-term disaster reduction. A review was carried out of historical storms that had a significant impact on nine sensitive European sites including wave exposure, tidal regime and socio-economic pressures. Monitoring over the course of one year to collect new data sets enabled us to develop and test numerical models of storm-induced morphological changes, linking wave and surge forecasting models to set-up a real-time warning system and to implement its usage within Civil Protection agencies. The project enabled the conception of Storms Impact Indicators (SIIs) with defined threshold for the identification of major morphological changes and flooding associated risks.

In parallel, the FP7 SIM.COAST project ("Numerical Simulation Tools for Protection of Coasts against Flooding and Erosion" - http://www.simcoast.eu/) contributed to improved process understanding, new knowledge, methods, new and improved numerical tools, resulting in decision support systems serving decision making at protection of coasts against flooding and erosion. Support to decision makers in improving co-ordination of coastal erosion and surface water flood risk – strengthens emergency planning arrangements.

The FP7 THESEUS project ("Innovative coastal technologies for safer European coasts in a changing climate" - http://www.theseusproject.eu/) developed a systematic approach to delivering both a low-risk coast for human use and healthy habitats for evolving coastal zones subject to multiple change factors. Innovative combined mitigation and adaptation technologies included ecologically based mitigation measures (such as restoration and/or creation of habitats), hydromorphodynamic techniques (such as wave energy converters, sediment reservoirs, multi-purpose structures and overtop-resistant dikes), actions to reduce the impact on society and economy (such as promotion of risk awareness or spatial planning) and GIS-based software to support defense planning. Eight study sites across Europe were selected, with specific attention Eight study sites across Europe were selected, with specific attention to the most vulnerable coastal environments such as deltas, estuaries and wetlands, where many large cities and industrial areas are located.

Catastrophic events such as the Xynthia event in France (February 2010) highlighted research needs in the prevention and preparedness of such extreme events, which were reflected in two major projects, namely: the FP7 PEARL project ("Preparing for Extreme And Rare events in coastal regions" - http://www.pearl-fp7.eu) developed more sustainable risk management solutions for coastal communities focusing on present and projected extreme hydrometeorological events. Seven case studies from across the EU were designed to develop a holistic risk reduction framework to identify multi-stressor risk assessment, risk cascading processes and strengthen risk governance by enabling an active role for key actors. Development of novel technologies and methods improved the early warning process and its components, building a pan-European knowledge base gathering real case studies and demonstrations of best practice across the EU to support capacity development for the delivery of cost-effective risk-reduction plans. Additionally, the project provided an interface to relevant ongoing tsunami work: it plugged into global databases, early warning systems and processes at WMO, and contributed to community building, development of guidelines and communication. Besides, the FP7 RISC-

KIT ("Resilience-Increasing Strategies for Coasts" - http://www.risckit.eu/np4/ home.html) developed ready-to-use methods, tools and management approaches to reduce risk and increase resilience to low-frequency, high-impact hydrometeorological events. Open-source and free-ware were developed to assess present and future hot spot areas of coastal risk due to multi-hazards, as well as high-resolution Early Warning and Decision Support System (EWS/DSS) for use on these hot spots (with a scale of 10's of km) and web-based management guide offering innovative, cost-effective, ecosystem-based DRR measures. Testing toolkits using data collected on 10 diverse case study sites were deployed along each of Europe's regional seas and one international site.

1.6. Science–policy interfacing

1.6.1. *Linking different actors*

The management of disaster risks such as coastal storms and floods is overseen by a number of international, EU and national policies covering various sectors and operational features such as preparedness, prevention, detection, surveillance, response and recovery. A wide range of research and technological developments, as well as capacity-building and training projects, are striving to support the implementation of these policies. However, the complexity of the policy framework and the wide variety of research, capacity-building and training initiatives often lead to a lack of awareness about policies and project outputs by the users, namely, policymakers, scientists, industry/SMEs and practitioners, e.g. civil protection units, medical emergency services along with police departments. Highly fragmented information often leads to poor awareness of policy requirements by research and industry communities along with poor transfer of research results to policy and stakeholder communities.

1.6.2. *Governance and knowledge transfer*

Several levels of governance need to be considered: (1) a "horizontal" level in the framework of which interactions among research, industry, policymakers and practitioners are established in a coordinated way at different scales, i.e. EU, national and regional. (2) A "vertical" level which establishes operational links between the EU, national and regional levels through appropriate information relays, synergies and demonstration activities.

The different levels are illustrated in Figure 1.2 and deal with, in particular:

1) Horizontally:

– *Science to science:* sharing information and developing interactions among H2020 research projects dealing with coastal risk and floods to develop a critical mass and reduce fragmentation, and bring tools/technologies to the market through links with industrial stakeholders. Projects should, in principle, respond to topics which are generally based on well-defined policy hooks (in this case, environmental and climate policies). Hence, we might expect that projects supporting common policy goals will establish synergies, which is rarely the case without a push from the European Commission owing to various considerations (intellectual property rights in particular). Sharing information and developing interactions on a regular basis should become a common practice.

– *Policy to policy*: policy interactions in the light of implementation needs, and establishing links with EU Member States. While International and EU policies are developed in close consultation between different sectors, in practice, few interactions take place at the implementation level between sectors within the Member States (e.g. climatologists working with hydrologists for policy implementation-related actions). This is partly due to insufficient sharing of information and joint actions.

– *Science to policy*: formatting/translation of research information in a way which is tailor-made to policymakers and ultimately user's needs, responding to well-specified technical challenges. This is directly linked to the above, with the requirement for the scientific community to format/translate research information in a way which is tailor-made to policy applications, basically responding to well-specified technical challenges.

– *Policy to science*: identification of research needs from policymakers, stakeholders and practitioners on the short- to long term and communication of these needs to be taken into account in research programming, development and implementation. An essential component of the policy to science interaction is the capacity for policymakers to identify research needs on the short to long term and communicate these needs in anticipation to the research community so that programming, research development and implementation can match the policytimeline (e.g. access to the scientific state-of-the-art, short-term research/capacity building, longer term research goals, and pre- and conormative research).

2) Vertically :

– *International/EU to National*: in the research sector, interactions through H2020 consortia; in the policy sector, interactions through Committees representing Member States and stakeholders, working out appropriate relays to national authorities and stakeholders based on well-formatted information. At international/EU level, policies are elaborated by relevant organizations (e.g. Sendai Framework for Action and European Commission for climate /environment-related EU policies). The links to the National level take place through Committees in which Member States are represented. There is a need to ensure that these

Committees are informed on similar grounds about science and policy developments.

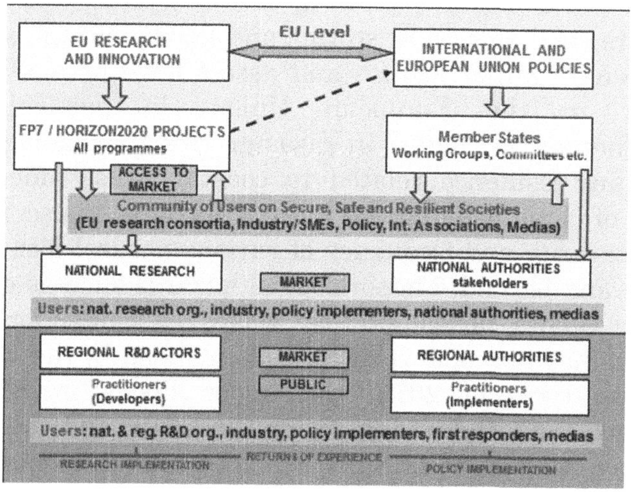

Figure 1.2. *Different levels of governance. For a color version of this figure, see www.iste.co.uk/quevauviller/storms.zip*

– *National to Regional / Local*: information is communicated through interactions with regional research partners and regional authorities as well as practitioner's networks and associations. Once representatives of the Member State's Committee are informed, it is to be expected that appropriate communication with regional / local implementers will then take place under the MS responsibility. This also requires a level of coordination, which depends upon the willingness and capacity of each Member State. This level of interaction is less well defined than the EU level because of different settings within the different Member States.

– *Regional to National / EU*: return of experiences from either practitioners involved in EU-funded projects or practitioners informed via national channels to the EU level.

1.6.3. *Networking needs*

In a world facing a growing risk of hydrometeorological disasters such as coastal storms and floods and the security of citizens – infrastructure and assets have become a high priority in the European Union, in particular for strengthening capacities in disaster risk management and improving resilience related to coastal risks. Indeed, the impact of climate change on natural hazards has seen a rise in the severity and frequency of various natural disasters in Europe and beyond. Meteorological hazards such as extreme weather events, floods and heat waves, as well as forest and wildfires have become recurrent phenomena in the EU. The Xynthia storm of 2010, the major floods in Southern Germany and neighboring regions in 2013, and the deadly heat wave which struck Europe in 2003 are a few examples.

Coastal risks management involves various communities covering research, policy and operational actors (including industry/SMEs, first responders, civil protection units, decision makers, etc.), all of which have specificities but share common features regarding the overall risk management cycle (preparedness/prevention, detection/surveillance and response/recovery) and the need to ensure a proper transfer (and implementation) of research outputs to "users".

This diversity of actors requires that the dissemination and communication of project results be tailor-made to different sectors, while bearing in mind that the common goal is to ensure that "solutions" resulting from research will reach users (often regional implementers, first responders, civil protection units, SMEs, individuals, etc.) in a timely and relevant manner and be translated into "useful & used operational tools", hence contributing also to the European economy through improved competitiveness. The high number of research projects and the lack of "interfacing" mechanism make it difficult to efficiently reach this goal.

1.6.4. *Who are the users of research?*

Fields concerned by coastal risks, including security, safety and resilience for societies are themselves scattered into different disciplines and sectors. In other words, we will distinguish five main categories of actors: (1) policymakers; (2) scientists; (3) industry (including SMEs); (4) training and operational units; and (5) NGOs and general public:

1) Policymakers and stakeholders:

– at the international level, UN bodies are closely working with the EU in the field of disaster risk reduction (UN-ISDR), environment protection (UNEP), etc.

– at the EU level, the main policies concerned with coastal risk management cover civil protection, environment and climate action;

– at the Member State's level, Ministries and Agencies in charge of Civil Protection, Environment, Research and Industry, as well as Agencies and Regional Authorities, are concerned;

– often working at the interface between policy and science, various stakeholders are involved in bridging interests of different communities, e.g. consultancy companies;

2) Scientists:

– Coastal risk-related research involves a range of scientific disciplines which have to interact, ensure complementarity and build interdisciplinary networks;

– Different types of scientists are to be considered (universities, research institutes and research units linked to ministries or agencies);

3) Industry (including SMEs):

– Many industry branches and stakeholders are involved in the areas of civil protection, environment and climate.

4) Practitioners:

– First responders, i.e. fire brigades, emergency services, civil protection units, water/flood management, etc., as well as decision makers (at national or regional levels).

– Training centers for first responders and command control centers.

5) NGOs and general public:

– NGOs, Civil Society Organizations, public at large, education (schools) and training.

While some of the above actors in categories 1, 2 and 3 are used to participate in international meetings, this is less frequent for SMEs (in category 3) and even less for actors in categories 4 and 5. New ways must be found to ensure that information may freely circulate "horizontally" as well as "vertically" in order to fertilize all project deliverables while, at the same time, maturing them to the final operational phase (also called "usefulness & use") by end-users, and integrating them into appropriate policy implementation and development.

1.6.5. *Building-up of a "Community of Users"*

The large span of projects leads to a huge dispersion of resources as no mechanism is presently in place to establish a common platform to exchange information of public character, boost awareness, transfer of relevant research projects to relevant users (and to industrial/SMEs share- and stake-holders) and make them "useful and used". In addition, efforts have to be made to better address users' needs which will be reflected into possible inputs to (EU and national) research programming. The need for a sharing platform led to the idea of developing a Community of Users along the principle shown in Figure 1.3.

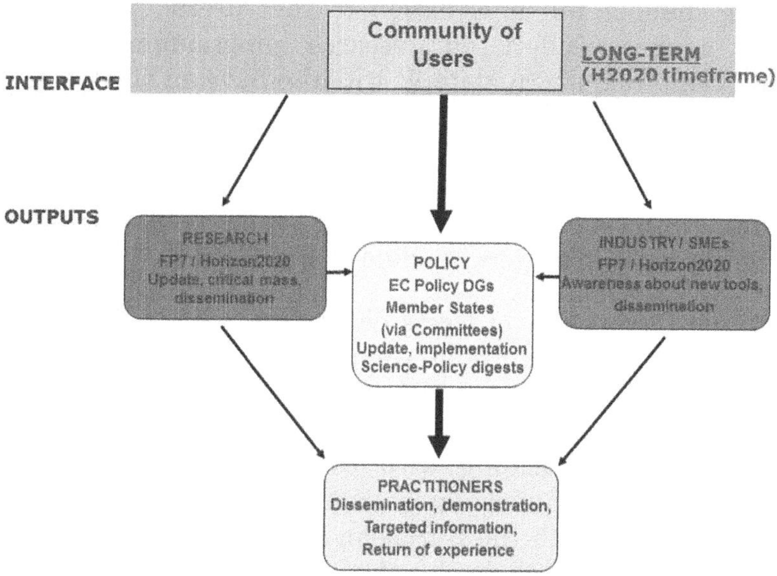

Figure 1.3. *Principles of the Community of Users. For a color version of this figure, see www.iste.co.uk/quevauviller/storms.zip*

The concept of "Community of Users" (CoU) is closely linked to the needs for a better coordination of information exchanges of general nature through a visible platform. An initiative has been launched by DG HOME (DG Migration and Home Affairs) since 2013, the objectives of which is to boost transfer of research outputs to relevant users and facilitate sharing of information among different actors. The objective is to create a mechanism involving different levels (EU, national and regional) by which the different actors, and primarily the "users", will be able to trace back information and experiences issued from research, capacity-building and training projects, giving them the opportunity to identify and contact the appropriate party at the right time to get the feedback that they need via a CoU-dedicated website. Regular information exchanges and debates orchestrated by the Community of Users will enable us to better channel the information to the "users", which will have a direct effect on research programming, policy implementation and update. It will also have an effect on the involvement of end-users at various levels, e.g. in steering committees of Horizon 2020 projects, consortia and cater links between research projects and capacity-building/ training initiatives, e.g. making links with training programmes and centers, module exercises, etc.

If the Community of Users develops as expected, it has a potential to become a useful, complementary, and supporting group on research-related activities to EU policies, including policies relevant to coastal risks (not duplicating existing advisory groups dealing with policy implementation but rather channeling information about research outputs) in the framework of which the European Commission with the EU Member States (through the policy and programme committees). The EU Agencies, Intergovernmental Agencies, International Organizations and the wide range of sectors concerned (research, industry, practitioners) will cooperate for boosting implementation of research outputs, including their usability for policy implementation in the Member States (through information given to relevant existing committees and advisory groups). This will, in addition, have the capacity of returns of experiences from Industry and practitioners to the EU level, and enable us to identify potential technologies, tools and methods in order to support their access to the market.

The Community of Users, along with a related initiative, namely, the Disaster Risk Management Knowledge Centre (DRMKC) coordinated by the EC Joint Research Centre, will enable us to better visualize/identify research (and on the long-term capacity-building and education) projects related to different themes relevant to safety, security and resilience. While this network is progressively establishing "horizontal" dialogues and helping interactions among different disciplines and actors, it will not have the capacity to create operational links with users at large without dedicated thematic networks (referred to as "Communities of practice") in Figure 1.4.

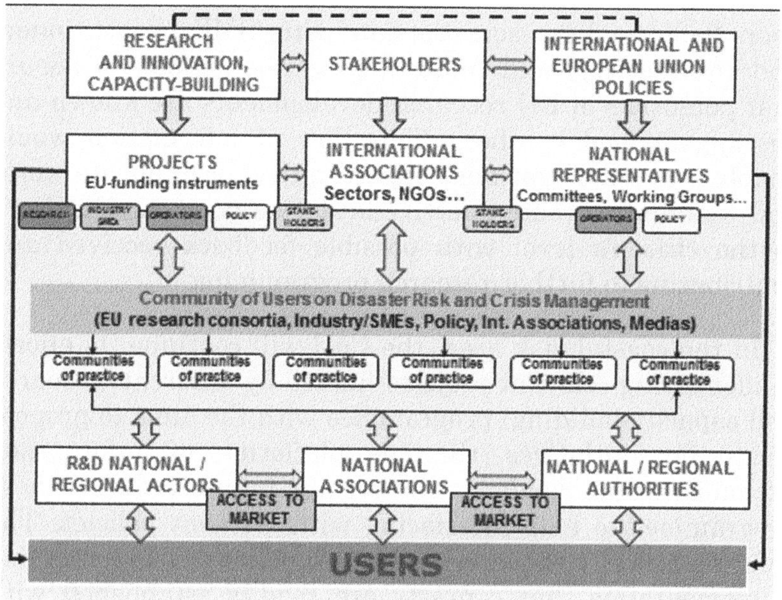

Figure 1.4. *Linking CoU to Communities of practice. For a color version of this figure, see www.iste.co.uk/quevauviller/storms.zip*

This need for "vertical" transfer of information from the EU to the national and the regional levels could be fulfilled by connecting the CoU to appropriate expert networks or communities, either existing or to be developed, which would play the role of knowledge integrating and "translating" bodies at European levels, with the mission – in support and connect with MS authorities – to effectively relay research outputs (e.g. new tools or technologies, methods, etc.) to appropriate users at national, regional and even local levels. This process of pulling EU research outputs to users, i.e. transforming these outputs into outcome, can only be possible through an effective partnership with users. In other words, if the CoU provides on a regular basis information on new tools/technologies or other research information, different "communities of practice" might format this information to address different categories of users (policymakers, scientists, industry/SMEs, practitioners and civil society) and undertake *ad-hoc* actions to ensure that potentials of EU research developments are known and possibly applied by them. This flow of information would enable that we do not miss opportunities (or duplicate work) and would also create effective bridges among the EU down to the citizen's level with possible feedback received and contributing to further research programming.

In the coastal risk area, the CoU will continue its efforts in identifying relevant projects funded by different (research and capacity-building) programmes with the aims to propose clustering initiatives through platforms of information exchanges. Stakeholders will continue to interact with these programmes to help interfacing with relevant policies. The CoU is naturally not interfering with policy development and implementation, but contacts are readily established with different policy bodies, enabling us to inform users about possible updates and helping research information to be efficiently disseminated to policy actors. The "Community of practice" needs to be activated to relay *ad-hoc* information to users as shown in Figure 1.5.

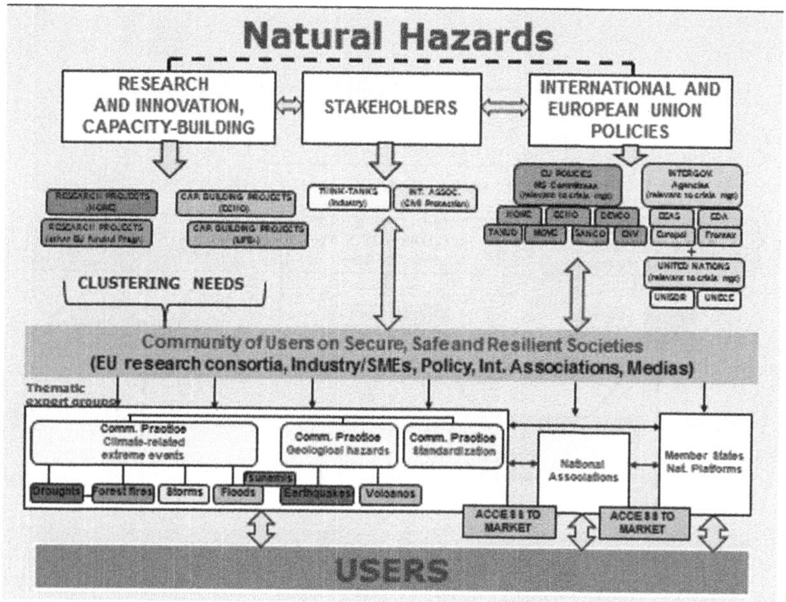

Figure 1.5. *Main actors in the Natural Hazards area. For a color version of this figure, see www.iste.co.uk/quevauviller/storms.zip*

Zooming into the picture (Figure 1.6), would imply that each Community of practice gets a comprehensive overview of leading projects in their area (research, capacity-building and training/education) and help bringing these projects. Interfacing among research and innovation and other actors in the industry and policy areas should be facilitated by stakeholder expert/ working groups with a mediating role, i.e. able to translate/format the information to target specially different users (e.g. specific technology information addressed to industry, support to a specific policy action with reference to the appropriate regulation, etc.). In bridging the different "worlds", there is a greater chance that users will get better, channeled information as the knowledge base would in principle become consolidated and made known to a wide range of different actors.

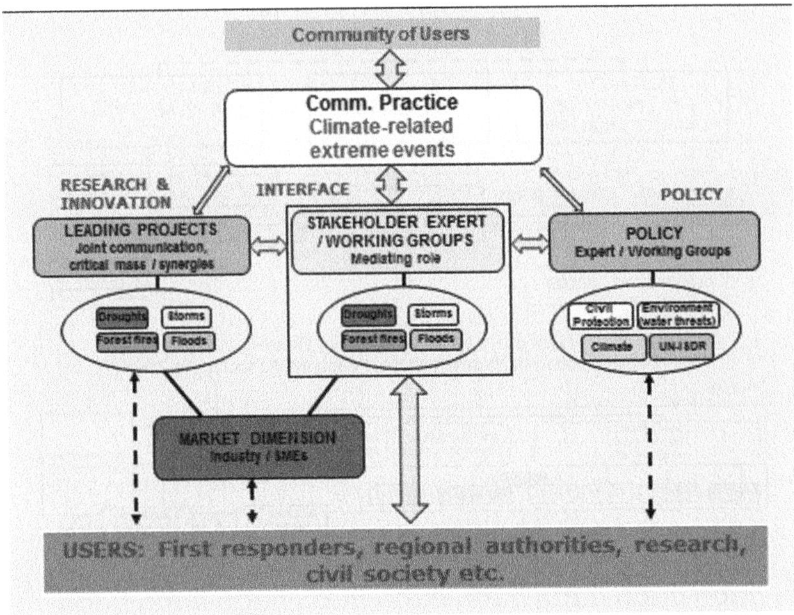

Figure 1.6. *Channeling information in the Natural Hazards area. For a color version of this figure, see www.iste.co.uk/quevauviller/storms.zip*

In conclusion, the Community of Users has the vocation to act as a facilitating platform, creating links and dialogues among different actors/disciplines (the "horizontal level") and among different levels (from EU to local). Based on the present mapping, a similar architecture will be used to develop a website which will facilitate information searches (not repeating what is readily in place but rather providing paths helping users to more easily find information per theme/areas). This mapping will be complemented on a regular basis (annually) for H2020 and other projects, and the CoU will pursue the organization of gathering events to consolidate a culture of exchanges at EU level for the sake of improved safety, security and resilience of our societies.

1.7. Conclusions

Most policies dealing with Disaster Risk and Crisis Management have established operational links with research. While interactions among research and policies are high on the policy agenda, much remains to be done to improve the way information flows from the different communities involved in implementation of both research outputs and policies. This includes capitalizing on past research and enhancing cooperation among EU Member States organizations. The complexity of the security sector arises from the wide variety of actors involved and the lack of coordination mechanism at EU and national level regarding the transfer of information and their actual use by implementers and decision makers. The need for enhanced coordination and information sharing forms the basis of the Community of Users on Safe, Secure and Resilient Societies described in this chapter.

Prior to developing a Community of Users (based on existing communities which are presently fragmented) with the view of improving science–policy–industry–operator's links in the context of Horizon 2020, it was essential to *understand the architecture* of the research framework and how it interacted with various policy technical/scientific challenges. This was the subject of the mapping which should not be regarded as an impact assessment (i.e. no analysis was done about the actual impact and the use of research outputs on policies) but rather as a means to better understand the complex science–policy working environment at EU and national levels and propose a mechanism to *streamline information flows and transfer in the future* [EUR 17]. The analytical value of the document stands for the "matrix" established between research and science, i.e. a factual image of the present situation. For the time being, it does not go as far as analyzing the real outputs of research regarding policy implementation but complements the work of the Commission's Disaster Risk Management Knowledge Centre (DRMKC), which intends to improve science-based services and analysis, the use and uptake of research and operational knowledge, as well as to advance science and technology in DRM.

What is the way ahead? Several objectives will be pursued, from the short- to the long term, which are described in details in the above referred report [EUR 17]. Besides the technical objectives and the coordination of a better information exchange system, the Community of Users on the long term has the capacity to rise sharing of experiences among different actors involved in disaster risk and crisis management (including coastal risks), with possible initiatives leading to synergies in the EU and beyond.

As a concluding remark, it should be recalled that policy orientations rely on scientific evidence. In this respect, the efficient use of science represents an increasing challenge for the scientific and policy-making community, the private sector, NGOs, citizens' associations and professional organizations. The need to improve the role that science plays in environmental policy-making has been widely debated over the last few years; in particular, the need to ensure better linkages between policy needs and research programmes, with enhanced coordination regarding programme planning, project selection and management, and mechanisms for knowledge transfer to ensure that outputs from research programmes really do contribute to policy development, implementation and review. This issue has been discussed in depth in the water sector at European Union level for more than 5 years, underlining the need to develop a conceptual framework for a science–policy interface related to water which would gather together various initiatives and knowledge [QUE 10] which is illustrated in Figure 1.7, showing the necessary links between research recommendations or tools and "users" (policymakers, stakeholders and water managers) and the need to ensure a "memory" of scientific information (facilitated by various dedicated websites), demonstration of the applicability of the research and dissemination through appropriate communication and translation of the scientific information. This closes the present chapter in underlining that the bridge between policy and research is non-trivial and deserves more attention from all actors concerned.

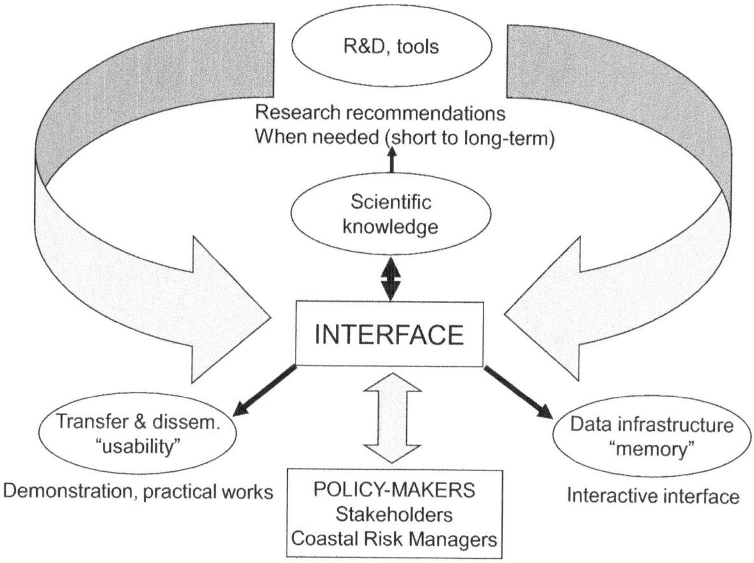

Figure 1.7. *Needs of interface between science and policy (adapted from [QUE 11])*

2

Techniques for the Assessment of Coastal Storm Risk

2.1. Introduction

As with many natural phenomena such as earthquakes, droughts or floods, storms are relatively unpredictable. Indeed, they vary considerably in duration and intensity, and usually occur very irregularly in space and time. However, using statistical methods and modeling, or through the use of measuring instruments and tools of observation, it is possible to provide relatively precise information on where and when a given storm is likely to occur, as well as to predict some of its parameters such as magnitude and duration. Considering the occurrence of coastal storms and their characteristics, it is finally possible to assess the risks, estimating the exposure of coastal sites and vulnerability of economic, natural and human assets. Storm Xynthia, which raged in Europe in 2010, and caused numerous dike breaches in the coastal area around La Faute-Sur-Mer (France), reminded Europeans of the importance of protecting people against the occurrence of extreme events and the need to develop a risk culture in order to prevent further disasters.

Wind has a major influence on the risk of erosion and, in the most extreme cases, on flooding in coastal areas. Its characteristics directly influence the sea state, specifically the height of the waves which is one of the main factors controlling coastal erosion. Indeed, the height of offshore waves depends on the wind speed, duration of action and the sea-extension over which it blows (fetch). However, near the coast, other factors – such as the slope of nearshore seabed – also have important effects on wave height, causing phenomena such as wave refractions, shoaling and breaking. Taking into account the characteristics of the waves approaching from the open sea and the local geomorphology, we can predict the transformation of the waves propagating toward the coast, the impact of the breakers on the shore and the amount of energy discharged by them.

Thus, the geomorphological study of beaches requires knowledge of aspects of hydro- and morphodynamics. In addition to the formation of waves, which are in part responsible for coastal erosion, storms can also cause an anomalous elevation of sea level: this is the phenomenon normally known as storm surge. It results from contributions from wind, wave set-up, as well as the inverse barometric effect due to the atmospheric low pressure. The surge contribution to anomalous water levels is defined as the positive difference between the observed and the predicted astronomical tidal tide (Figure 2.1).

Storm surge is a temporary and local elevation of sea level generally associated with a low atmospheric pressure system and strong winds blowing toward the coast. Associated with high tidal coefficients, this phenomenon produces a significant risk of erosion and/or marine flooding of coastal areas (Figure 2.2).

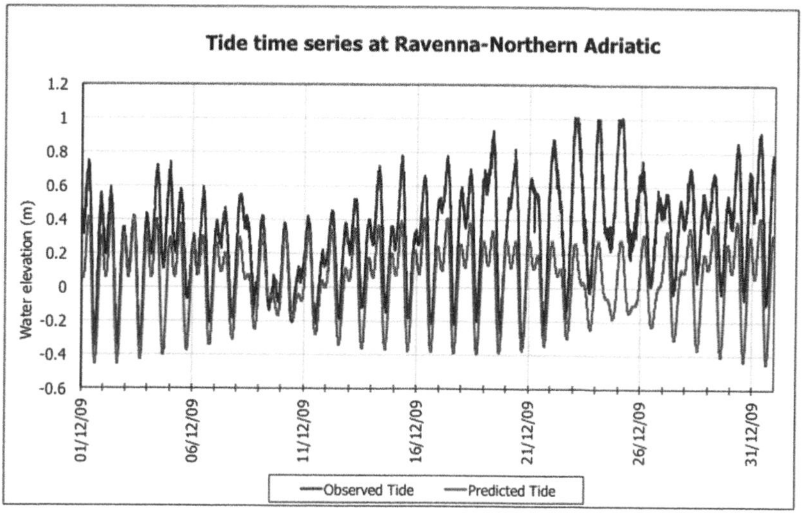

Figure 2.1. *Example of a storm surge record from a tidal station located at Ravenna, along the northern Adriatic coast of Italy. Notice how the atmospheric contribution can double the height of the astronomical tide. Data elaboration by M. Masina. For a color version of this figure, see www.iste.co.uk/quevauviller/storms.zip*

The assessment of coastal vulnerability is not based solely on physical factors as described above. Vulnerability also depends on economic, social and environmental factors (see Figure 1.2). For example, coastal communities are particularly vulnerable to erosion and flooding when a considerable number of residential infrastructures, trade activities and transport utilities are close to the beach, in low-lying areas at risk from coastal flooding. Similarly, sparsely inhabited coastal areas characterized by significant biodiversity – particularly wetlands, deltas and estuaries – are also threatened by floods and rising sea level, and their ecosystem value may be at risk. Coastal protection structures have well-defined costs for construction and maintenance, but do not always provide concrete solutions to the risk of erosion and marine submersion. Sometimes, coastal protection structures may even amplify the effects of storm surges and waves. By pushing the sand offshore during storms, they may cause a lowering of beach levels, or even a complete disappearance of the beach in front of them. The morphodynamics of the beaches behind parallel protection structures becomes peculiar, and care must be given to locate the structures at the correct distance from the shoreline, as well as with the correct spacing if several items have to be placed. Ultimately, the natural ability of coastal systems to absorb wave energy is weakened as the structure either withstands the waves or fails, and the risk of flooding may increase if the structure's crest is not adequately high above the maximum surge level [GAL 14].

Figure 2.2. *Example of storm-induced beach erosion and flooding of commercial activities located at Lido delle Nazioni (Ferrara), along the northern Adriatic coast of Italy. Picture taken on 5 February 2015 (courtesy of L. Perini). For a color version of this figure, see www.iste.co.uk/quevauviller/storms.zip*

These coastal hazards are likely to be exacerbated by climate change. According to the IPCC (Intergovernmental Panel on Climate Change), climate change and an increase in sea level will result in an amplification of coastal flooding due to storms, as well as an increase in coastal erosion, threatening almost all coastal megacities [NIC 11]. Indeed, although no real consensus has been reached on the extent of the phenomenon and the affected regions, a number of studies have identified regions of the world which, according to climate scenarios, are subject to an increase in the magnitude of storm waves [NIC 11]. The inclusion of different global warming scenarios and rising sea level therefore appears to be necessary to predict the long-term impact of storms on coastal areas [VOU 16].

2.2. Definition of coastal risk

In the context of natural hazards, risk can be generally defined as the probability of harmful consequences, or expected losses (deaths, socio-economic and environmental damage) resulting from interactions between hazard and vulnerability (equation [2.1]). Thus, the risk is usually represented as follows:

$$\text{Risk} = \text{hazard x vulnerability} \qquad [2.1]$$

where:

– the hazard can be defined as the probability that a particular event (threat) will occur within a given time period (return period level);

– the vulnerability can be defined as the degree of loss of an item or group of items located in the area affected by the hazard.

However, some disciplines also include the concept of exposure, to specifically refer to the physical aspects of vulnerability. In the context of coastal vulnerability, these physical aspects include the location of infrastructure, population density levels, design and material of building, coastal erosion magnitude, rate of rising sea level, etc. More generally speaking, vulnerability is not only limited to physical forcing, but it also depends on the conditions determined by social, economic and environmental factors that may influence the degree of exposure of coastal areas to a given natural hazard. Thus, the impact of storms on coastal environments should be determined whilst taking these factors into account, in order to assess the vulnerability and therefore the risks of coastal regions and populations.

2.3. Hazard time series in Europe and beyond

For a given geographic location, storm hazard is defined by its intensity as well as by its temporal occurrence (time and frequency). The latter can be quantitatively estimated, for example, by computing the return period of storm events (or probability of occurrence), either on a purely statistical basis or by considering recorded impacts on coastal areas.

An historical analysis of storms from a physical viewpoint can be achieved through the collection of meteorological (or hydrodynamic) data over time. Access to this type of data, such as the occurrence of storms, as well as the records of their effects (e.g. on coastal property and infrastructure), is an essential element in the assessment of coastal risks, and for predicting future risk. The concept of looking at historical impact datasets to reconstruct the severity of storms has been refired by Baart *et al.* [BAA 11], who reconstructed estimates of 1 in 10,000 year return period of extreme flooding in the Netherlands. Although several databases have been developed at the local or regional level – for example, [LAM 91] for the North Sea, or [PFI 10] for Portugal, northern and western Europe – the MICORE project (http://www.micore.eu), having actively contributed to the archiving of historical data on storms on the coasts of

Europe, was the first to produce a Europe-wide assessment [CIA 11a].

In the context of climate change, variability in storm patterns in Europe and rising sea levels are likely to cause an increase in coastal flooding, as well as an increase in erosion, thus necessatating specific adaptation strategies [EUR 10, EUR 14]. Indeed, climate simulations using the IS92a scenarios, A2 and B2 of the IPCC Special Report on Emissions Scenarios (SRES), predict an increase in wind speed and intensity of storms in the northeast Atlantic over the 2010–2030 period. Moreover, a decrease in the intensity of the winds in the eastern Mediterranean, and some localized increases in storms in parts of the Adriatic, Aegean and Black Sea have been identified [SOT 06]. Regarding the North Sea, some studies using different combinations of general circulation models (GCMs), regional climate models (RCM) and regional hydrodynamic models for storm surges (and waves) were able to estimate a change in the future storm regime of the same order as natural climate variability [EUR 14]. However, the analysis of meteorological and hydrodynamic variables could provide trends of observed change: this is the case, for example, in the North Adriatic and the southern Baltic Sea, where the frequency of storm surges has increased [CIA 11b]. This hypothesis was supported by the results of statistical analysis conducted under the MICORE project [CIA 11b]. One of the main difficulties encountered by the MICORE project in assessing the changing storm patterns was accessing time series of meteorological and hydrodynamic data that were sufficiently long and representative. Furthermore, these databases are not always publicly available at national levels, and some use restrictions which they sometimes apply. However, based on available data from recent decades, after data quality control, a synthesis of the general trends in the evolution of storms systems in Europe (duration, intensity and frequency) was established by Ciavola and Jimenez

[CIA 13] for different coastal regions. This synthesis was primarily an experimental approach based on a set of meteorological and hydrodynamic data (or proxies), such as waves, winds and storm surges. In many cases, these datasets did not extend beyond a few years or, at most, a few decades. When the data available only cover a few years, it is difficult to discriminate longer term trends over interannual variability driven by general circulation phenomena like the NAO and the El Niño/La Niña. Furthermore, very few long-term datasets are available for beach erosion to allow us to compare physical forcing with observed shoreline erosion trends. In a recent paper, Barnard *et al.* [BAR 15] managed to examine 48 beaches across the Pacific Ocean and correlated the signals of El Niño/Southern Oscillation using data between 1979 and 2012 that described wave climate, local water levels and coastal change. Arguably, this is the longest dataset of this kind for open ocean basins. Such a correlation for enclosed seas like the Mediterranean, the Baltic or the Black Sea may not display such a clear signal.

Analysing the effect of climate change on coastal storms and their impacts on coastal environments through the estimation of return periods of extreme events, generally requires the use of a long time series of meteorological and hydrodynamic data. However, we should remember that instrumental observations of waves only became available after the end of World War II. Tide measurements sometimes have longer time series, which date back to the 19th Century, but attention should be paid to changes in the reference level of the gauge. In Figure 2.3, a long-term time series of surges is presented that shows an increase with time partially due to local subsidence.

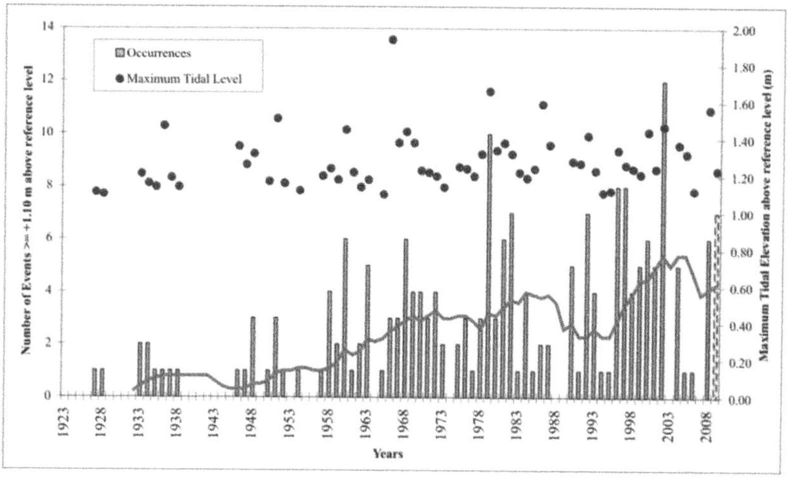

Figure 2.3. *Characteristics of the storm surges recorded at the tide gauge of Venezia (Punta della Salute) in the period 1923–2008. Data from Comune di Venezia-Centro Maree. The line is the 10-year moving average of the events observed every year. The reference level is local and corresponds to MSL at the tide gauge (1897 datum). For a color version of this figure, see www.iste.co.uk/quevauviller/storms.zip*

One solution to the problem of limited time series, is to extend existing databases using models based on historical simulations (or "hindcast"), particularly for waves, such as the HIPOCAS database or other databases as ICOADS (http://icoads.noaa.gov). The HIPOCAS database (Hindcast of Dynamic Processes of the Ocean and Coastal Areas of Europe) is a set of data at the European level, established on the basis of historical (1958–2001) simulations of meteorological and hydrodynamic data such as wind, waves and the sea level (http://www.mar.ist. utl.pt/hipocas).

Forward forecasting of the storm hazard mainly depends on the time scale being considered. Storm predictions can essentially be done at two temporal scales: short term (hours to days) and medium- to long term (months to years).

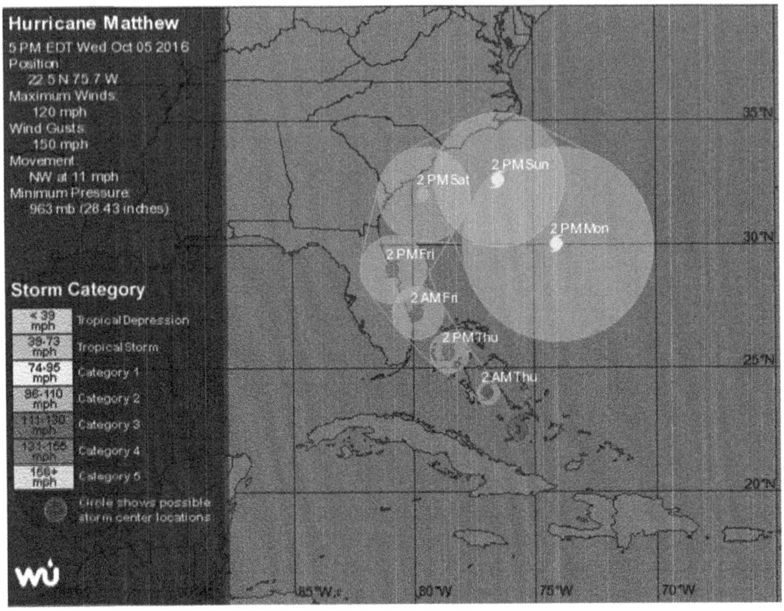

Figure 2.4. *Example of track forecasting of hurricane Matthew, which hit the Caribbean, Florida and the Southeastern coast of the USA in October 2016. The weather system reached peak strength of Category 5 (https://www.wunderground.com). For a color version of this figure, see www.iste.co.uk/quevauviller/storms.zip*

2.3.1. *Short term: hours to days*

In the short term, predicted coastal storm hazard is mainly based on weather forecasts. The sea state is determined accordingly (as wave height is mainly dependent on the characteristics of the wind). Short-term forecasts require means of predictions and observations such as those used by meteorological services. However, forecasting storms, their intensity and their extension remain relatively difficult. This applies to tropical storms as well as extra-tropical systems, the latter being more complicated as these weather systems rapidly change. Recent examples where the forecast was problematic are Superstorm Sandy along the western coast of the USA [TOL 13], Xynthia cyclone along the western European coastlines [BER 12] and cyclone Hayan in the Philippines [MAS 15]. As the most recent example at the time of writing, we could cite Hurricane Matthew (Figure 2.4), a strong tropical cyclone over the Atlantic Ocean, which became the longest lasting Category 4 hurricane on record, and was the first Category 5 hurricane since Hurricane Felix in 2007.

Matthew formed from a tropical wave that moved off the African coast on 22 September 2016, following a westward track until developing into a tropical storm while situated just to the east of the Leeward Islands on 28 September 2016. A day later, it became a hurricane while west of the Leeward Islands and rapidly strengthened into a Category 5 hurricane. It impacted Haiti, Jamaica, Cuba the Dominican Republic, the Bahamas, the Eastern United States including the states of Florida, Georgia, South Carolina and North Carolina. It was downgraded to a Category 3 when it hit the US coastline, but massive evacuation procedures were still activated.

2.3.2. *Medium- and long term*

In the longer term, forecasting the characteristics of coastal storms (frequency, duration and intensity) involves other assessment techniques including those that determine the evolution of the storm system. For a given region, it is useful to consult the archives and the time series of meteorological and hydrodynamic data to assess the best general trend and the analogies with previous events (Figure 2.5). When the time series are incomplete, it is possible to use models based on retrospective simulations of weather or hydrodynamic parameters. Finally, climate models can be used to understand teleconnections between large-scale ocean circulation and wave climates [LIO 05] and to predict future wave climate changes [AND 07].

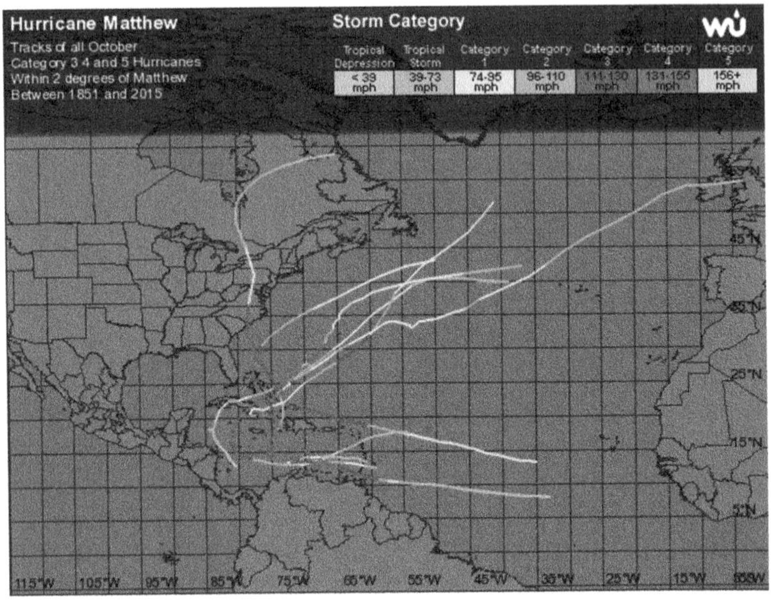

Figure 2.5. *History (1851–2015) of October hurricanes with location comparable to Matthew. It is immediately noticeable that the historical record testifies how not all events die out in the central Atlantic, with one event actually making it across the ocean and hitting northern Ireland. (https://www.wunderground.com). For a color version of this figure, see www.iste.co.uk/quevauviller/storms.zip*

2.4. Evaluation of coastal vulnerability

2.4.1. *Evaluation on the basis of critical thresholds*

One of the issues in assessing coastal vulnerability during storms is to link the events characterized by morphological changes and major damage to the hydrodynamic forcing. However, the precise definition of storms beyond thresholds that produce significant morphological changes or significant damage to coastal structures is not systematically described in the scientific literature. Often, we only encounter a defined wave height limit above which we consider a storm to occur, with or without significant morphological changes or damage [CIA 11b]. Nevertheless, under the project

MICORE, critical thresholds of indicators related to hydrodynamic forcing were identified [CIA 12]. The theoretical approach to obtaining these critical thresholds in the MICORE project was based on the collection of data relating to significant morphological changes recorded in recent decades. The collection and use of historical sources, topographic and bathymetric data, aerial photographs, reports, etc., has proven to be an effective means of recording storm impacts. Storm impacts were identified on the basis of their hydrodynamic characteristics and analyzed to define both the morphological and damaging thresholds. For example, for the entire studied coastal areas, wave heights or wave energy were used as criteria for the definition of these thresholds. In addition, the majority of the studied coastal areas used storm surge water levels as criteria; for some coastal areas, the duration of the storm, return periods, direction, wave run-up or tide levels were used as additional criteria. Finally, data on erosion, damage from storms, coastal flooding, overwashing, dune erosion and impacts on coastal buildings were used as additional criteria to characterize the storm threshold that triggers the onset of such events [CIA 11b].

Note, however, that there is no universal approach here. This technique for obtaining thresholds can be recommended when, for a given region, enough observations and data relating to storms and damage exist. Indeed, the defined thresholds are, in most cases, based on field data. Further information on critical storm thresholds can be found in [CIA 12].

2.4.2. *Coastal risk maps*

The procedure for assessing the vulnerability of coasts to storm surges usually requires the completion of risk maps. For example, in response to the recurring problem along US coastlines, the Federal Emergencies Management Agency (FEMA) played an important role in identifying vulnerable areas through the provision of maps used by insurance companies to provide cover against flooding. These maps helped citizens to obtain insurance policies against floods and accurately reflected the risks for a particular region. The procedure adopted by [FEM 03] for the identification of risk areas considers factors such as hydrodynamic and meteorological forcing (waves, tides and storm surges), as well as resilience (human or natural).

At the European level, the methodology suggested by the FLOODsite project (sixth Framework Program of the European Union) – to obtain the coastal flood risk maps – underlines, again, the importance of the interaction between the beach profile, wave characteristics and the level of the run-up – namely, the maximum elevation reached by the waves during storms. This project was the precursor of the Directive 2007/60/EC of the European Parliament and the Council of the European Union of 23 October 2007 (see Chapter 1). This requests that member states map the geographical areas subject to flooding, under different scenarios based on return periods of events. According to this Directive, a flood risk map for each scenario should include the following factors: (1) the extent of the flood, (2) the depth of water and, if appropriate, (3) the current speed. Nevertheless, given the specific characteristics of coastal flooding, hazard maps may indicate not only the extent of the areas likely to be flooded, but also areas along the coast that are vulnerable to other coastal hazards such as beach and dune erosion significant enough to cause flooding of the hinterland. More generally, the process of assessing coastal vulnerability to storm surges requires the completion of maps based on several hydrodynamic and geomorphological factors such as waves, run-up, erosion, dune erosion and overtopping. All these factors contribute to the Total Water Level, which is the level of inundation to be used in the flood maps. However, the risk of flooding is generally based on the return periods of extreme storm events. An example of a coastal flood map is given in Figure 2.6.

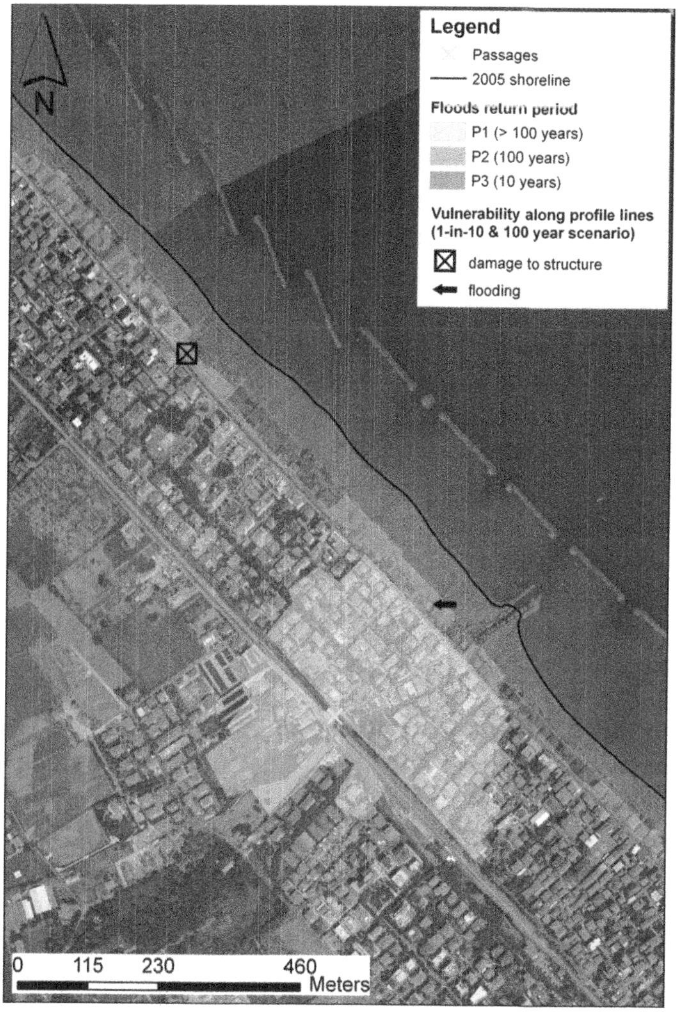

Figure 2.6. *Viserba, Rimini province, Italy. Flood hazard map prepared by the Emilia-Romagna Region for marine floods with different return periods. The symbols represent the vulnerability typologies of the 10 and 100 years return period scenarios. The yellow cross represents the location of a low-lying passage which acts as a funnel for inland flooding [PER 16]. For a color version of this figure, see www.iste.co.uk/quevauviller/storms.zip*

2.4.3. *Topographic and bathymetric surveys*

Beach geomorphology depends on many factors such as tides, beach slope, waves, currents, sediment grain size, the erosion rate of the beach itself and nearby dunes, as well as the impact of storms. Under extreme hydrometeorological events, the beach profile is generally heavily altered and may not recover to its previous state during fair-weather conditions. In fact, the risk of dune erosion or submersion of the backshore and damage to coastal infrastructure depends strongly on the configuration of the beach immediately before an event, e.g. a beach severely eroded during a storm may lead to flooding under successive storms. An accurate analysis of beach variability requires bathymetric and topographic measurements possibly carried out as repeated surveys on a regular basis. These bathymetric and topographic surveys are usually performed *in situ* using the GPS (or differential GPS) technique and single or multibeam echosounders. This technique usually involves taking measurements along transects spaced at regular intervals along the beach but on some occasions, e.g. in a flat beach profile, the use of an ATV vehicle allows full coverage of large areas.

Compared to other topographic methods widely used in the past (e.g. line and rod method, Emery method, theodolites and total stations), DGPS has several advantages for beach surveying: (1) it allows very high precision measurements along transects; (2) surveying can be quickly performed after extreme storms to determine their immediate effects (e.g. loss in beach elevation, water levels reached by the storm, etc.); (3) the entire beach profile can be measured, even extending below low water levels if the operator uses waders; (4) and it is a job that can be performed by a single person [MAS 00]. The main disadvantages are related to the difficulties of (1) covering very large areas; (2) choosing a sufficiently representative transect of the surrounding beach; (3) and taking readings during favorable weather conditions (daylight, good weather, etc.). However, even the simplest bathymetric surveying technique (e.g. single-beam echo-sounding) needs access to a floating device (e.g. a small vessel, a jet-sky or a kayak) and may be carried out under harsh wave conditions, which may be dangerous for the operator. The use of a Multibeam Survey System instead provides almost complete coverage of the nearshore seabed, generating a terrain model that is less prone to artifact produced through interpolation (Figure 2.7). The only drawback of MSS is the relatively high cost compared to traditional single beam as well as the high sensitivity to quality control of the surveyed data. Frequent bar checks (e.g. speed of sound checks) are needed, and pitch and roll corrections must be accurate.

Figure 2.7. *Digital terrain model of the seabed north of the Port of Ravenna (Italy) obtained with a Multibeam Survey System. Notice the smoothness of the contours compared to a traditional interpolated survey. Dataset courtesy of ENI S.p.a, background photo Google Earth. For a color version of this figure, see www.iste.co.uk/quevauviller/storms.zip*

Topo-bathymetric surveys may also be done through the use of advanced remote sensing technologies such as LIDAR (Light Detection and Ranging), radar [RUE 02] or video monitoring techniques like ARGUS [HOL 07] or other low-cost video systems [HAR 14] Traditional aereo-photogrammetric techniques based on cameras mounted on piloted vectors (planes and helicopters) are now being replaced by the use of drones [TUR 16], which have the advantage of easy mobilization, low cost and are easy to launch from any location There are still technological restrictions such as a limited payload for commercial drones, battery duration and stability of simple platforms (e.g. quadcopters vs. hesacopters), especially under strong wind conditions, when fixed-wing vehicles must be preferred.

LIDAR technology is an active remote sensing tool for the execution of high resolution surveying [IRI 98]. It is normally based on an airborne laser using a narrow light beam to measure distances, which has been used worldwide since the mid-1990s for mapping coastal changes [MIT 03]. Indeed, the analysis of Digital Terrain Models obtained by LIDAR, especially before and after an extreme storm, can be used to assess coastal erosion magnitudes, and more generally, geomorphological changes caused by the storm. This makes it possible to predict future impacts, including taking appropriate coastal protection measures. For example, [CIA 07] used this technology to study, the impact of storm surges along the coast of Emilia Romagna (Italy). Using LIDAR airborne surveys performed at an interval of approximately 1 year (2003–2004) – with the second flight conducted after a major storm – the geomorphic impact induced by a storm with a return period of 25 years could be evaluated. Arguably, this work was the first of its kind on Italian territory.

Following this pioneering work, within the framework of the BEACHMED-e project (http://www.beachmed.it), local Italian government agencies like ARPA Emilia-Romagna (the Regional Agency for Environmental Protection), charged with assessment of coastal engineering projects, developed methods for a topo-bathymetric characterization of the regional coastline using LIDAR, which helped to compare this method with commonly used technologies like traditional topo-bathymetric survey. Similarly, in the Languedoc-Roussillon region of France where the coastline is experiencing significant sediment dynamics, the LIDAR technology was recently used for a bathymetric monitoring study at the regional level [HEU 08].

Generally, during LIDAR surveys for topographic measurements, a red laser is used, which cannot penetrate water. However, if a survey is properly planned, e.g. flying around Low Water Spring Tides, a good coverage of the intertidal zone can be provided. Bathymetry can instead be determined in the subtidal zone using a green laser by recording the backscatter from the surface of the water, an intense light reflection, followed by a lower return from the sea bottom. In case of exceptional water clarity, penetration can reach as much as 70 m [FIN 05]. A LIDAR obtained DTM provides a dataset that typically ranges from 0.5 points/square meter to 1 point per square meter. Topographic features like dune elevation, gaps in the dune ridge and preferential low-lying locations for water ingression during storms can be easily identified (Figure 2.8).

The plane flying the LIDAR system is often equipped with an integrated digital camera with high-resolution optics. There are in fact many advantages associated with the use of LIDAR data: it is an inexpensive and effective way to collect field data with high precision and a good spatial and vertical resolution [BRO 09] which is particularly useful for digital terrain modeling. LIDAR topographic surveys can generally produce datasets with a vertical precision of 10 cm [KLE 09]. Regarding bathymetric surveys in shallow water, a limiting factor is the water clarity. As a rule-of-thumb, penetration of the laser beam is in the order of three times the Secchi Disk visibility, so surveys in turbid water are not recommended [FIN 05]. Although the data collection proves to be advantageous in terms of cost, the acquisition cost of the LIDAR equipment is very high; furthermore, LIDAR cannot be used in case of cloud cover, fog or rain.

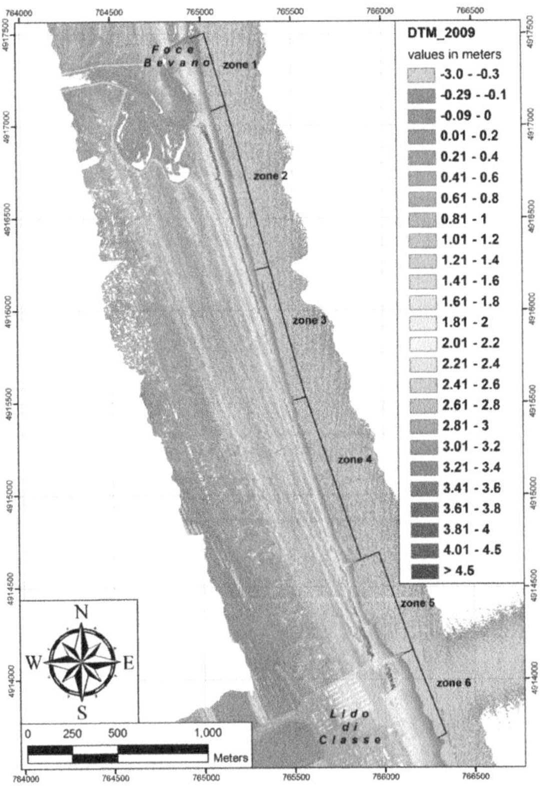

Figure 2.8. *Digital terrain model of a dune system north of Lido di Classe near Ravenna (Italy) obtained with a LIDAR flight. Notice how the dune ridges that testify historical coastal progradation can be identified. Notice also how the dune crests of Zone 2 and 5 are clearly higher than those located in the central part. For a color version of this figure, see www.iste.co.uk/quevauviller/storms.zip*

Radar technology works on the same principle as LIDAR technology, with the difference that the active sensor uses radio waves (microwaves) for measuring distances instead of light waves (laser scanning). This technique of airborne or satellite observation includes imaging tools such as ERS, ENVISAT, RADARSAT and SENTINEL using the SAR – Synthetic Aperture Radar. Bathymetry of shallow water can also be assessed by the use of radar. Indeed, the effect of bathymetric features is that they are visible on the SAR images through the current changes in radar signature of the ocean surface and the modulation of corresponding waves. Generally, the method of obtaining bathymetry from remotely sensed data is still under development and much more research is needed to use it at an operational capability level [MCI 06, PAC 15].

Finally, a video system like ARGUS is one of the technologies that, among others, can provide much-needed topo-bathymetric measurements on beaches. This technique allows video monitoring of coastal processes by producing high quality digital images [HOL 07]. Approximately 30 ARGUS stations and 120 cameras have been operational worldwide. The imaging is done continuously, typically every hour of the day, with a pixel resolution of 1 m to cm. The analysis of *Timex* pictures (time exposure) – created on the basis of the average of pixel intensities, collected once per second over a 10 min period – allows, after filtration, to identify and locate points of wave dissipation through persistent foam patterns. The location of the breaking waves (dissipation) can finally serve as proxy to locate submerged features. Note that while video-derived intertidal bathymetry measurement is quite an established technique [ALE 04, ARM 06, HOL 07], the only tools developed so far for bathymetry estimations are the Beach Wizard [VAN 08] and C-bathy tools [HOL 13, BER 16].

2.4.4. *Estimation of wave parameters*

Wave characteristics can be observed and studied using various tools and measuring instruments. Airborne altimeters or satellites with active sensors (e.g. radars) can provide data on the significant wave height, on the currents, and on the wind speed at the sea surface. This method can provide a valid alternative to wave and current measurements performed using fixed stations like buoys or Lagrangian measurements like drifters. The significant breaking wave height (H_b) is particularly important in the risk assessment of erosion and flooding associated with extreme weather and sea conditions. In a simple form, it was mathematically defined by [KOM 72] using the following equation:

$$H_b = 0.39\sqrt{g}\left(H_s T_p\right)^{2/5} \quad [2.2]$$

where g is the acceleration due to gravity, H_s is the significant offshore wave height and T_p is the peak wave period.

Moreover, wave power is also an important factor to consider in assessing forcing conditions during storms. The energy of a breaking wave (E_b) can be related to its height H_b, using the following equation according to linear wave theory:

$$E_b = (\rho\ g\ H_b^2)/8 \quad [2.3]$$

where ρ is the density of sea water (kg/m^3).

In European countries, there is unfortunately little long-term date available for measured waves. However, there are databases of historical simulations, although reliability remains limited in some seas with a peculiar oceanographic setting (e.g. the Northern Adriatic). That is why it is recommended that all European countries should contribute to the establishment of a European database of available wave measurements.

The characterization of waves for the evaluation and reduction of risks associated with erosion and coastal flooding can be done using hydrodynamic models for wave forecasting. In this context, the ECMWF version of the third-generation Wave Model Cycle 4 (WAMC4), developed by the WAMDI Group (1988), is one of the most popular and better tested wave models. The model predicts the wave height, wave period and directional fields [ALV 10]. The main usefulness of the model is its ability to characterize the sea state, especially during storms, but the model can also help to complete time series of measured data when they are incomplete. The study undertaken by Cieslikwicz and Paplinska-Swerpel [CIE 08] applied to the Baltic Sea has validated the WAM model by comparing the modeled time series with observed data. Comparisons between the model and data measured by buoys and satellite altimetry observations (Topex/Poseidon) showed that significant wave height tended to be over predicted during the storm peaks.

Unlike the model WAM, specifically designed for oceanic applications, the SWAN (Simulating Waves Nearshore) model, another third-generation wave model, was designed to model the waves in shallow water. This model is particularly suited to the study of waves in coastal areas as it is able to consider processes induced by shallow-water wave transformation, including wave breaking [BOO 99].

2.4.5. *Wave set-up*

The quantification of the extent of wave setup has been the subject of numerous studies over the years, among which is necessary to cite that of Bowen *et al.* [BOW 68] who suggest the following formula:

$$\frac{<\eta>}{H_b} = \gamma \qquad [2.4]$$

Dove $<\eta>$ is the setup, H_b is the breaking wave height and γ is the ratio between H, wave height in the surf zone (shallow water conditions and wave fronts parallel to the coast) and h, local water depth.

There were many subsequent reinterpretations of the analytical terms of this parameter: Guza and Thornton [GUZ 81] show a setup proportionality with offshore H_s, whereas Holman and Sallenger [HOL 85] emphasize the need to take into account the Iribarren (ξ) number. Finally, on dissipative beaches, Hanslow and Nielsen [NIE 91, HAN 93] confirmed that the slope of the beach is irrelevant for the purposes of calculating the maximum setup. The latter line of thought is also confirmed by Stockdon *et al.* [STO 06] whose formulations are in fact different for dissipative and intermediate-reflective beaches.

For the calculation of wave-induced set-up, it is useful to know the Iribarren number for each profile as, according to Stockdon *et al* [STO 06], the formulation of set-up is different depending on the type of morphodynamics associated with each beach. In particular for dissipative beaches ($\xi < 0.3$), the following applies:

$$< \eta > = 0.016 \ \sqrt{\text{Lmax Hmax}} \qquad [2.5]$$

While for intermediate and reflective beaches, the formula becomes:

$$< \eta > = 0.35 \, \beta \ \sqrt{\text{Lmax Hmax}} \qquad [2.6]$$

where:

Lmax is the maximum wavelength.

Hmax is the maximum wave height.

β is the beach slope.

2.4.6. *Wave run-up*

The run-up is defined as the maximum elevation reached by the wave uprush on the beach. It can be determined by measuring the highest level reached by water above the mean sea level, measured at the line of breaking waves, and is the result of two factors: (1) the set-up (i.e. the average elevation of the water on the shoreline, generated by the incident waves, basically a reasonably stationary process along several wave periods); (2) the swash (that is to say the vertical water fluctuations from the average level which are due to waves rushing back and forth along the beach slope). This is an oscillatory contribution and is partially related to the incident wave period, but also to longer term oscillations (the well-known "surf beat") that provide a secondary energy component. The accurate prediction of the run-up height (R) is essential for risk evaluation of the impact of storm waves on natural and man-made features. R can be mathematically formulated by combining the offshore wave parameters (height, period and wavelength) with the slope of the beach. Initially, Hunt [HUN 59] suggested the following relationship:

$$R = H_0 \; \xi \qquad [2.7]$$

where H_0 is the deep water significant wave height and ξ is the Iribarren number [IRI 49]. This can be defined as the beach "dynamic" slope, and is formulated as follows:

$$\xi = \tan\beta \,/\, (H_0 \,/\, L_0)^{1/2} \qquad [2.8]$$

where tanβ is the beach slope, H_0 is the offshore wave height and L_0 is the offshore wavelength. This parameter is a classic in the beach morphodynamics literature and was made popular among researchers with the seminal paper of [WRI 84].

Run-up is generally indicated as $R_{2\%}$ that represents the height exceeded by 2% of the highest water levels. This value is often used as a threshold value in the study of coastal vulnerability [SAL 00, SAL 03]. The $R_{2\%}$ parameter was initially set to determine the wave run-up on coastal protection structures and was then used to determine more generally maximum water levels on beaches. Its formulation depends on an empirical constant that is determined experimentally. For example, according to the formula of Battjes [BAT 71], the $R_{2\%}$ can be calculated as follows:

$$R_{2\%} = (C\xi)\ H_0 \quad [2.9]$$

where C is the constant to be determined. Many formulas have been used to calculate run-up in studies of coastal erosion and vulnerability. For example, Armaroli *et al.* [ARM 09] evaluated the vulnerability of beaches protected by breakwaters in Emilia-Romagna (Italy) using a re-written form of the original mathematical formula of Holman [HOL 86]:

$$R_{2\%} = (0.83\ \xi + 0.2)\ H_0 \quad [2.10]$$

Other methodologies for assessing coastal vulnerability have been established by combining run-up observations with processes like wave overtopping of dune ridges. A study by Vousdoukas *et al.* [VOU 12], for example, managed to make forecasts of overwash (overflow above low-lying dune ridges) using offshore wave observations and video monitoring, leading to swash and run-up quantification on Faro beach in Algarve (Portugal). Run-up datasets were generated from video images (timestack and sigma images) observing variations in pixel intensity along transversal transects covering the swash zone. All data relating to the run-up were first compared with existing parameterization (some involving parameters such as elevation of the sea and wind speed), and then with experimental and specific parameterization to that site in order to assess vulnerability to overwash along the coastline.

Recent studies [RUG 04] have pointed out that the morphodynamics conditions of the beach should be accounted for in the analytical formulation of the run-up. Holman and Sallenger's [HOL 85] solution seems to be appropriate for intermediate and reflective beaches, [GUZ 81] for moderately dissipative beaches, while Ruggiero *et al.* [RUG 01] and [STO 06] propose a range of different formulas, taking into account whether they have to be used on reflective or dissipative beaches (notice that [RUG 01] is based on data from both the east and west coasts of the United States, while the second is mainly derived from the analysis of data from the east coast). Stockdon *et al.*'s [STO 06] formula is possibly the most complex one in its general term:

$$R_2 = 1.1 \left\{ 0.35\, \beta_f \sqrt[2]{Lmax\ Hmax} + \left[\frac{Hmax\ Lmax \left(0.563\, \beta_f^2 + 0.004\right)^{1/2}}{2} \right] \right\} \quad [2.11]$$

However, this can be greatly simplified for dissipative beaches:

$$R_2 = 0.0043\ \sqrt{\mathrm{Lmax\ Hmax}} \quad [2.12]$$

It is necessary to specify that both methods already include the setup value, which is not necessarily added to the computation. However, it can also be noted that the first equation takes into account the beach slope (βf), which is not taken into account in the second one, because the authors considered it irrelevant in the case of dissipative and ultra dissipative beaches.

2.4.7. *Numerical models for beach dynamics*

Different computer models can be used to evaluate the morphodynamics changes experienced by a beach during storms. Here, we will focus on cross-shore profile models, as it is along this dimension that wave transformation plays an important role in generating beach erosion and dune/structure overtopping. The SBEACH model (Storm-induced BEAch CHange Model), developed by the U.S. Army Corps of Engineers (USACE), has proved to be an effective model for simulating the changes in beach profiles and dune erosion caused by storm surges and wind waves [LAR 04, DON 06]. The formation and movement of major morphological features of the beach profile such as beach ridges, swales and berms can be modeled. The model, being a cross-shore model, can only be applied if variability of the waves, currents and sediment transport along the coastline dimension can be neglected. The main factors responsible for sediment transport and change of beach profile in SBEACH are the role of breaking waves and changes in water level produced by storm surges, tides and wind.

The bidimensional numerical model XBeach (eXtreme Beach behavior), developed in collaboration between UNESCO-IHE, Deltares, Delft University of Technology and the University of Miami, is a model specifically designed to simulate the response of the beach to the impacts of extreme storm events [ROE 09]. The model includes, among other features, the breaking process, surf zone and swash zone dynamics. Moreover, it is able to model dune erosion and predict phenomena such as overwash and breaches in the barrier beach. Unlike some simulation models of the impact of extreme events on dune erosion, where their application assumes certain conditions – such as SBEACH, see previously – XBeach proves to be a more flexible model for measuring the impact of storms in more complex situations. For example, while some models assume uniform conditions along the coast, XBeach was developed so that the process variability along the coast is considered. The model gives detailed attention to long-wave processes (e.g. the InfraGravity Wave Band), which play an important role in generating base dune erosion. Finally, it helps to assess the impact of extreme events in situations where barrier islands protect the coast from storm impacts. Indeed, elevation, width and length of the barrier island, and the hydrodynamic conditions of the backshore must be taken into account to evaluate the morphodynamics response of this kind of system.

2.4.8. *Development of vulnerability zones*

In coastal areas, damage and built capital-related losses are related to the location of buildings, particularly their distance from the beach. West *et al.* [WES 01], for example, included this concept in a probabilistic approach in which the probability of damage to infrastructure decreases linearly with distance from the beach. Therefore, the definition of zones of vulnerability is fundamental for estimating the economic losses caused by coastal storms. In this context, identification of parameters such as the erosion rate of the dunes or the retreat rate of the coastline during storms can prove to be very useful. This approach allowed researchers working on the MICORE project to identify areas of vulnerability, especially as they developed Storm Impact Indicators (SIIs) [CIA 11a]. This method of estimating damage based on the definition of vulnerable zones can be applied based on a diachronic analysis of aerial or satellite images – see e.g. [HON 10].

The evaluation methods for losses caused by storm surge are important because they help to assess the socio-economic impact (actual or potential) of on environmental assets that characterize some coastal regions to extreme events. This type of assessment provides insight into future risks and aids in decision making on adaptation and risk reduction to be implemented. This assessment of incurred losses (tangible or intangible) is done by different approaches, which generally consider physical data such as weather-marine data and data relating to coastal infrastructure, socio-economic and environmental costs, the economy and the natural and human heritage.

2.4.9. *Development of damage curves*

Damage curves are functions that put the hazard parameters (such as wind speed or water depth) in relation with physical damage, social or economic loss. They have been used in particular by the Federal Agency for emergencies in the United States (FEMA) in a standard economic model called HAZUS-MH (Multi-Hazard Loss Estimation). These functions are used in different models (coupled with GIS applications) that estimate the potential losses caused by extreme storms (Hurricane Wind HAZUS-MH Model), as well as losses caused by floods (HAZUS-MH Flood Model) [SCA 06, SCH 06, VIC 06a, VIC 06]. The estimation of losses is, however, mainly based on physical damage to the structures and buildings and therefore includes the direct economic losses. However, losses due to the interruption of production processes can also be calculated (e.g. based on the loss of revenue). The method developed by FEMA is very accurate but requires relatively large efforts related to data collection.

2.4.10. *Input-output economic model*

The input-output model is a model used to estimate the indirect costs caused by the interruption of economic activities resulting from a "shock" as the occurrence of a natural disaster. This model assesses how a shock caused, for example, by a storm surge, could affect the economic system by the changes in supply and demand that it induces. Specifically, the model assesses changes in the interrelationships between different economic actors such as industry and consumers. The basic principle assumes that industry requires inputs produced by other industries, while production of this industry will serve as inputs to other economic sectors. The method of determining the flow of goods and services between different industries is applied in order to predict how the economy will react to an external shock and how it will evolve over time – see, for example [HAL 08] for the evaluation of the economic cost of Hurricane Katrina. This method, however, requires a major effort to use it and for the calibration of data sources when it comes to adapting the model to a geographical area and a specific period.

2.4.11. *Climate change scenario and predicted losses*

Finally, climate change is a variable that must be integrated into various models for assessing coastal vulnerability through the use of different scenarios of rising sea level and/or changes in storm characteristics. As for the loss assessment methods, climate models that incorporate the concept of risk variables may include physical, environmental and socio-economic variables.

Climate change, especially the increase in sea levels, is an important factor to consider when assessing the long-term impact of extreme storm events and marine submersion. In this context, it is necessary to integrate socio-economic and environmental variables and scenarios resulting from rising sea levels. The SimCLIM models DIVA [BAR 08, HIN 10, SPEC, 16] allow us to integrate these different variables and can therefore be used to inform decision makers on the effects of rising sea levels in areas where resources and population are closely related [IGL 10]. These types of models are proving to be very useful for national and international authorities when assessing vulnerability and adaptive capacity for reducing the impacts of climate change.

2.5. Toward disaster risk reduction

Risk management is often seen as a systematic process to evaluate and implement strategies necessary for reducing the impacts of natural hazards on society and to improve control of the affected communities' capabilities. It is often presented as a cycle comprising: (1) a disaster response action during or immediately after its occurrence, (2) a recovery process, and (3) a reduction in risk, which essentially seeks to prevent or reduce the damage caused [KRE 14]. It is the latter category that applies measures such as the construction of coastal defense structures or Early Warning System implementation to limit the impacts of storm surges. Examples of Early Warning Systems are briefly described in this chapter because they play a special role in decreasing the impact of these events on coastal populations. Monitoring of beaches and the sea state and the use of impact indicators are needed in the design of warning systems related to coastal storms and risk of marine submersion. Evaluation on the basis of critical thresholds makes it possible, for example, to decide when to trigger the alert and plan an appropriate response strategy.

2.5.1. *Monitoring the storm impact*

Beach monitoring for the protection of coastal areas from storm impact at a European scale was conducted for the first time by the MICORE project: a monitoring network was implemented in various parts of Europe across nine countries covering all regional seas. Implementation of beach monitoring activities was done over a 1–2-year period, according to the sites, with the aim of: (1) collecting new data on bathymetry and topography using advanced technology (LIDAR, ARGUS, Radar and DGPS), and simultaneously measuring forcing agents (wind and waves, tides and storm surges) that triggered the events, and (2) mapping the impact of storms on living and non-living resources. In addition, monitoring datasets also helped to validate the XBeach code that at the time of the monitoring (2008–2009) just being developed and used to predict morphological changes induced by extreme storm events on the US coastline. In addition, the calibrated model was implemented inside a series of prototype Early Warning Systems capable of rapidly predicting the occurrence of erosion or flooding.

To progress in coastal risk assessment methods, the MICORE project developed impact indicators or Storm Impact Indicators (SIIs). These impact indicators were made applicable to a variety of coasts (natural or artificial), based on European case studies, considering various degrees of wave energy, tidal range and socio-economic characteristics. The indicators were initially designed for use by civil protection agencies to help in managing the evacuation of people, sending staff to monitor dike breaches in vulnerable places and implementing various emergency measures, such as the installation of sandbags and temporary dikes, or define security zones where people could find refuge in case of flooding [CIA 11a, CIA11b].

Among the SIIs, or proxies used in the development of them, there were for example: beach erosion, flooding, dune overtopping, and the location of properties and infrastructures potentially subjected to damage [CIA 11a,

CIA 11b]. According to the peculiarities of the study sites, different SIIs were applied. For example, for the region of Emilia-Romagna (Italy), two impact indicators were defined: (1) the width of a safety corridor (Safe Corridor Width-SCW) between the dunes and the water level to allow passage (2) and the distance between the buildings and the shoreline (Building Waterline Distance-BWD) [HAR 12, HAR 16, HAR 13]. These indicators can be used for the construction of risk maps, which can then be presented for different return periods of storm events. The indicator SCW, for example, can be used to visualize the actual degree of risk in the event of a storm, based on a GIS map. Moreover, it can be used to prevent losses in the urban area behind the beach. One should remember that in Mediterranean countries, hotels, shops and roads are literally built on the backshore and it is easy to imagine that inundation could take place during sustained periods of high water levels produced by atmospheric, tidal and wave processes. If a numerical beach model is run in an operational mode, using the waves and tides forecast, the SCW indicator can be determined in advance and used to define critical thresholds for the implementation of an alert system (see next section). In a more general way, for the various sites studied by the MICORE project, the SIIs quantification took place mainly in the application of the XBeach model. In detail, the thresholds for impact were defined on the basis of scenarios of storm categories, historical data from past storms, new weather, and hydro- and morphodynamic measurements collected as part of the project. Table 2.1 summarizes the framework of the MICORE concepts and quantitative models or derivatives of measures that were used for the quantification of the SIIs.

Risk management approach	Quantitative concept	Strategic objective	Operational procedures
Monitoring of dunes and dykes (under extreme forcing conditions)	Probability map indicating the most prone flooded coastal areas, developed unambiguously identifying the places at risk and the duration of the flooding events	To ensure an efficient response to threats during major storms	From the comparison between the calculated water levels and a predetermined acceptable level, a map showing the probability of flooding can be built and used as a basis for the deployment of dike surveillance staff
Protection of private properties on the beach (under extreme forcing conditions)	Risk maps with expected economic damage in the coastal zone (including homes, businesses, etc.)	Economically optimal protection of the greatest possible number of properties during storm conditions	Use of available protective measures to enhance local resilience at critical points to minimize the economic damage
Protection of private properties on the beach (under moderate forcing conditions)	Run-up time series extracted from a beach morphodynamic models or simple empirical notations	Support entrepreneurs of recreational infrastructure by preventing the damage caused by storms	When the SII indicates an impact on private property, the owners of the properties should be alerted as soon as possible
Protection against coastal erosion – conservation of natural areas (Directive 92/43 / EEC of the EU Council)	Run-up level, orthogonal and longitudinal extent of maximum sea flood limit	Ensure lasting security of the natural heritage	Protection of areas with deployment of temporary protection measures
Bathing safety (under average forcing conditions)	Spatial and temporal map of the areas that are considered dangerous for swimming, covering at least the most popular regions	Prevent injuries and casualties among beach goers under storm conditions	Reporting of conditions and hazardous locations: evacuation or rescue of people in areas at risk

Table 2.1. *Overview of the framework for Storm Impact Indicators (SIIs) used in MICORE to develop a standardized approach for the operational management of coastal risks. The quantitative concepts used are the SIIs (gray column) associated with the strategic objectives of the project and response procedures*

2.5.2. *Operational Early Warning Systems for surges*

Several EWSs for storm surges exist across Europe and in other parts of the world [BER 12]. Although meteorological warning systems have often been successfully developed for predicting severe weather events such as tropical cyclones and hurricanes, EWS for storm surges are not always effective, or they can even be absent from areas that suffer from this process. In reality, only a few regions in the world seem to have operational and efficient EWSs for storm surges.

Early Warning Systems for extreme hydrometeorological events and associated storm surges need to be operational for low-lying regions subject to hurricanes or tropical cyclones such as India or Bangladesh, where most of the land is less than 10 m above sea level [PAU 09]. Here, human losses following these events can be in the order of thousands of people for each cyclone season.

However, many other regions are particularly vulnerable to severe cyclones and induced storm surges. Seven tropical cyclone "basins" where storms occur on a regular basis have been defined as follows:

1) Atlantic basin (including the North Atlantic Ocean, the Gulf of Mexico and the Caribbean Sea);

2) Northeast Pacific basin (from Mexico to about the dateline);

3) Northwest Pacific basin (from the dateline to Asia including the South China Sea);

4) North Indian basin (including the Bay of Bengal and the Arabian Sea);

5) Southwest Indian basin;

6) Southeast Indian/Australian basin;

7) Australian/Southwest Pacific basin.

If we take the Pacific Ocean as an example, it is divided into basins, which have a Regional Specialized Meteorological Centre (RSMC) with the responsibility of issuing tropical weather outlook and tropical cyclone advisories for the benefit of the countries in the World Meteorological Organisation (WMO).

Warnings of extreme weather events are generally communicated by national weather services. For example, the India Meteorological Department issues cyclone warnings for the Indian coast. In the United States, the Storm Surge Unit of the National Hurricane Centre is able to provide accurate real-time surge forecasts during tropical storm events and to support coastal community preparedness and resiliency. By using the SLOSH Model (Sea, Lake, and Overland Surges from Hurricanes), the Storm Surge Unit also models and predicts storm surge vulnerability over a large area that includes the United States Atlantic and Gulf coasts, Hawaii, Puerto Rico, the U.S. Virgin Islands and the Bahamas. More information about SLOSH can been found in [GLA 09] and [MEL 10].

Regarding the USA coastline, according to Artigas *et al.* [ART 09], fairly accurate warnings of sea surges do exist; however, few spatially explicit warning systems are currently operational. A study conducted by a State regional planning agency (New Jersey Meadowlands Commission, (NJMC) integrates existing surge warnings from NOAA, sensor readings from tide gates, and spatial information about the Meadowlands Estuary of New Jersey into a real-time ocean surge warning system, see also [LEE 11]. At least 3 h in advance of high water, the system produces detailed maps of 14 Municipalities in the estuary showing which properties are most likely to be affected. For details about the components and the mechanism of the flood warning system see [ART 09].

In addition, the USGS plays an important role for the implementation of storm surge warning systems see, e.g., [CAL 10]. Indeed, it has developed a "mobile network of rapidly deployable water-level and barometric-pressure sensors to better observe and document the timing, magnitude, and duration of hurricane-induced storm surge dynamics as they make landfall and interact with coastal features" [CAL 10]. The National Weather Service (NWS) also has an important role since it is able to issue warnings for severe storms and coastal floods.

In Europe, coastlines are also vulnerable to coastal flooding and meteorological services issuing warnings for storm surges also exist. Table 2.2 provides some examples of storm surge warning systems in Europe.

In Europe, many models have been developed for the simulation of storm surges in the last decade. Recently, operational systems, such as the storm tide warning service in the UK, the real time storm surge forecasting system in the Netherlands, and the surge warning system for the North Sea, have been developed and improved in Western Europe. An interesting review of operational storm surge modeling at the European level is given by De Vries *et al.* [DEV 95] who report on the performance of the storm surge forecasting models of IFREMER (France), MUMM (Belgium), KNMI (The Netherlands), POL (United Kingdom) and the University of Athens and Aristotle University of Thessaloniki (UA/AUT, Greece) in a few different situations.

Region	Service	Hazard definition in the local language	Risk level designation definition in the local language
France	*Météo France*	*Vagues-submersion*	*Une vigilance absolue s'impose* *Soyez très vigilant* *Soyez attentif* *Pas de vigilance particulière*
United Kingdom	*Environment Agency*	*Flood*	*Severe Flood Warning* *Flood Warning* *Flood Alert*
Spain	*Agencia Estatal de Meteorología*	*Fenómenos costeros*	*Riesgo Extremo* *Riesgo Importante* *Riesgo* *Sin Riesgo*
Italy	*Servizio Meteorologico Aeronautica Militare*	*Eventi costieri*	*Molto pericoloso* *Pericoloso* *Potenzialmente pericoloso* *Nessun Avviso*
Europe	*EUMETNET*	*Coastal Event*	*The weather is very dangerous* *The weather is dangerous* *The weather is potentially dangerous* *No particular awareness of the weather is required*

Table 2.2. *Examples of storm surge warnings issued by national weather services in Europe. The risk-level designations are in the local language used for the warnings. Notice that for Italy, the agency in charge of weather forecasting is the Air Force and does not currently issue a national warning for coastal floods but a general warning for "coastal events". Here, the warning for Italy comes from the interface of the European MeteoAlarm Service. For a color version of this table, see www.iste.co.uk/quevauviller/storms.zip*

The following sections give some additional information about flood warnings in coastal areas for Northern Europe, the Atlantic coasts (France, Spain and Portugal) and the Mediterranean Sea. The reader is also referred to [PRO 83, FLA 00, WOL 09] for a review of operational systems used for real-time prediction of tides, surges and waves in northwest Europe. Additionally, Kolen *et al.* [KOL 09] provide a comparison between Dutch and German preparation practice for flooding, especially in early warning applications, and the use of flooding scenarios and evacuation strategies.

In Germany, the forecasts of imminent storm tides are made by the Hamburg storm tide warning service (WADI), part of the Ministry of Economics and Labour. Regarding mitigation measures, the project XTREMRISK was initiated in order to improve the understanding of risk-related issues due to extreme storm surges and enabled the quantification of the flood risk for two pilot sites at the open coast [BUR 10].

In the Netherlands, the storm surge warning service (SVSD), in close cooperation with the Royal Netherlands Meteorological Institute (KNMI), is responsible for sea water-level forecasting [VER 05]. Forecasts are made to support the national storm surge flood warning system. The surge is predicted by using a numerical hydrodynamic model, the Dutch Continental Shelf Model (DCSM), which is also used for the coasts of other countries along the North Sea. In the Netherlands, there are also local operational flood warning systems such as the Warning service Dikes IJsselmeer area (WDIJ) that was created to issue warnings, due to dangerous storm situations, for the dike authorities when a risk of flooding is expected to occur, see [CLA 99, DIN 09].

In the United Kingdom, the Environment Agency (EA) is responsible for issuing coastal flood warnings in England and Wales, based on wind, wave and storm surge forecasts produced by the MetOffice [FLO 10] through the Storm Tide Forecasting Service. It is worth mentioning the UK Coastal Monitoring and Forecasting (UKCMF), a partnership of public bodies which is working together to provide a comprehensive coastal flood forecasting service. Its strategy is described in the Environment Agency reports of 2009 and 2011. In Scotland, the Scottish Environment Protection Agency (SEPA) has developed a flood warning system to provide local authorities and emergency services with up to 24 h warning of coastal flooding [KAY 05]. Regarding studies for warning systems in the UK, [BRO 10] presents a case study of combined wave and water levels under storm conditions using the models WAM and SWAN (see also [BRO 09] for the eastern Irish Sea). Likewise, Dawson *et al.* [DAW 11] present an agent-based model for risk-based flood incident management by considering flood-warning systems. The Environment Agency [ENV 11] reviews recent advances in weather forecasting capability in the United Kingdom and their implications for increasing the lead-time with which flood warnings can be issued. More information on flood warnings in the UK is available in [WOO 04], while Horsburgh *et al.* [HOR 09] provide a clear explanation of the link between operational surge forecasting and decision-making procedures. Regarding mitigation strategies in the United Kingdom, Bradbury *et al* [BRA 05] describe the use of coastal monitoring data to develop strategic and operational beach management plans in southern England.

In Denmark, the Operational Oceanography Division has the duty of issuing storm surge warnings for the Danish Waters [BUC 05]. However, three governmental institutions operate approximately 40 tide gauges, all providing data in real-time: the DMI, the Danish Coastal Authority and the Royal Danish Administration for Navigation and Hydrography. The purpose of this monitoring is primarily to support storm surge warning, navigational safety and coastal protection [FEN 08].

In Norway, the Marine Forecasting Centre at DNMI Bergen issues warnings when there is a risk of water level exceeding given criteria. Warnings are sent to responsible organizations, e.g. harbor authorities, local authorities and police stations at about 100 addresses along the coast, and to national and local radio and TV stations for inclusion in weather reports. The Norwegian Hydrographic Office (SKSV) operates a network of tide gauges, with real-time transmission of the data to DNMI [FLA 00].

In Estonia, a model has been developed (HIROMB-SMHI model) [LAG 11] for the production of sea-level forecasts for the Estonian coast. An automated high/low sea-level checking is performed for each forecast. Critical values are defined and when the forecast is out of the defined limits, an automated high/low sea-level warning message is sent to the users. For measurements of the sea level along the Estonian coast and the forecasts, go to Sea Level Information System) http://on-line.msi.ttu.ee/kaart.php?en.

In France, since 2011 Météo France has been in a position to deliver a specific weather warning related to coastal flooding. For each coastal Department, Météo France provides a color-coded warning (i.e. green, orange or red) based on the forecast sea level and wave heights [LUM 11]. The monitoring of sea levels in France is undertaken by the "Service Hydrographique et Océanographique de la Marine" (SHOM). Regarding the French Mediterranean coastline, the HYDROGUARD-Espadhom is a local system that has been applied in Thau Lagoon (France). Ayral *et al.* [AYR 11] give an overview of the automated system for the monitoring and management of coastal risks and water resources, which includes a submergence warning system (using meteorological storm surge forecasts).

In Spain, a storm surge prediction system called Nivmar was developed and predicts sea levels for Spanish coasts [FLA 00]. The model is used to predict the surge component driven by meteorological forcing only. Water levels are computed as the sum of surge and tide predicted using the harmonic method from analyses of data from tide gauges of the REDMAR network. For more information about Nivmar, see also [CAR 00, FAN 01]. For the Mediterranean Sea, models for prediction of coastal dynamics have been developed at a local scale for a number of applications. For example, Alvarez-Ellacuria *et al.* [ALV 10] present a nearshore wave and current operational forecasting system for the north-eastern part of Mallorca Island (western Mediterranean).

In Portugal, sea-level change (SLC) real-time monitoring has being developed. A software application named *MareVB* receives a 3-min stream input of sea-level height and a 10-min stream input of air-pressure. Based on a predicted tide model, the sea-level height is compared and analyzed, and storm-surge amplitude is determined, as well as the high-frequency oscillation due to the storm waves. The application is now running as a coastal hazard warning system, emailing automatic warnings in real-time to national authorities and to other institutions [ANT 11]. The ongoing development and application of a methodology for evaluating the risk of coastal flooding will enable the development of a system for flood forecasting and warning for coastal areas and ports in Portugal (see [RAP 11]).

In Italy, proper operational chains for sea-level predictions are only implemented for the Adriatic Sea. Bajo *et al.* and Bajo and Umgiesser [BAJ 07, BAJ 10] describe an operational surge forecast system based on a combination of a hydrodynamic model and an artificial neural network. The system runs at the Centre for Sea-Level forecasting and flood warnings of the Venice Municipality (Istituzione Centro Previsioni e Segnalazioni Maree (ICPSM)) and is focused on the prediction of the surge near Venice. The hydrodynamic model provides a 5-day forecast for the Mediterranean Sea. More information about the development of an integrated forecast system in the Northern Adriatic Sea can also be found in [BAR 02]. See also [LIO 06] for an operational prediction of storm surge in the northern Adriatic Sea.

2.5.3. *Operational Early Warning Systems for beach morphological changes*

Through the development of the Storm Impact Indicators, the MICORE project first established a prototype system able to alert coastal populations at risk in case of flooding and support emergency preparedness for the implementation of measures [CIA 11b]. The construction of risk maps was a first step in the establishment of new warning systems, but the real innovation was the prediction of the impact based on dynamically varying beach, no longer considered only as a surface over which water was overflowing during storms [CIA 11b]. Taking as an example the coastal areas located in Emilia-Romagna (Italy), operational risk maps were built based on several beach profiles at well-known hot-spots [HAR 13]. Here, the concept of morphological Early Warning Systems was created and, second, the indicator Safe Corridor Width (SCW) was chosen to predict coastal variations depending on weather and sea forecasts (wave and surge). In the present configuration, the system has been operational since 2011 and is used to predict when the beach is too narrow (that is to say below a predefined threshold value) to allow people to walk without any danger on the beach as a function of the severity of the predicted sea conditions. More specifically, the safe corridor is defined to be equal to the horizontal distance between the foot of the dune and the intersection between the total water level (calculated using XBeach) and the beach profile [HAR 16]. The dune foot is manually defined on each profile line and the initial position of the dune foot is used to calculate the SCW (although the model predicts erosion or destruction of the dune). Colors, used to define the level of risk, were then chosen as defined threshold values for the impact indicator in accordance with indications given by the local end users. The reference thresholds identify three levels of risk: no risk when the SCW is greater than 10 m (green); medium risk when the SCW is between 5 and 10 m (yellow); and high risk when the SCW is less than 5 m (red). An example of prediction of SCW is shown in Figure 2.9.

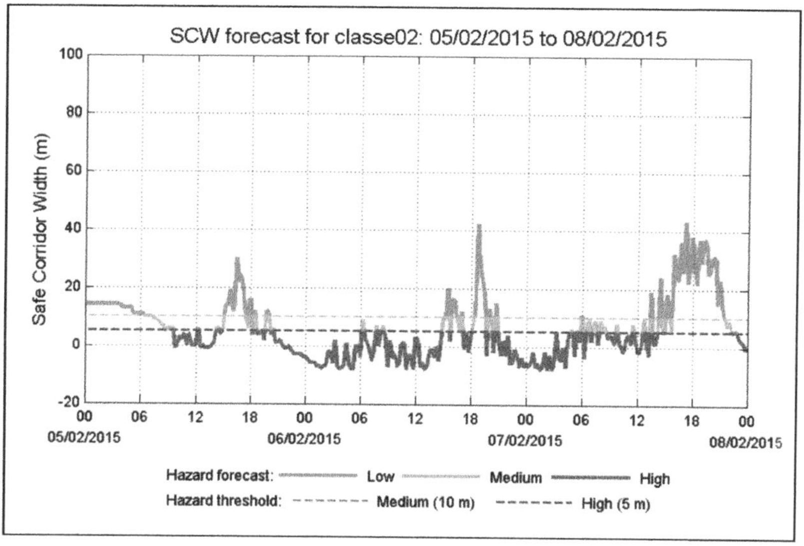

Figure 2.9. *Output of the Emilia-Romagna Early Warning System (http://geo.regione.emilia-romagna.it/schede/ews/) for the indicator Safe Corridor Width during a major stormy period occurred 5–7 February 2015. For a color version of this figure, see www.iste.co.uk/quevauviller/storms.zip*

Baart *et al.* [BAA 09] present a case study in the Netherlands where real-time forecasts of morphological storm impacts have been applied in the context of the MICORE Project (www.micore.eu). For the Sefton Coast (Northwest England), Esteves *et al.* [EST 12] have quantified thresholds for significant dune erosion in support of the establishment of an early warning system. For the coastal site of Dziwnow Spit (Poland), the Storm Impact Indicators were built on the basis of overwash, dune erosion and flooding of the beach [BUG 13, BUG 15]. For each of the studied coastal sites, the threshold values used to assess the level of risk were associated with green (for a low risk), orange (for an average risk) and red alert (for a high-risk level). The use of these colors in early warning systems, on the other hand, allows anyone to immediately identify where and when high-risk situations arise, in order to take the necessary preventive measures. In this case, an alert can be issued according to the prediction displayed on the hazard map and a signal can be placed on the ground so that people going to the beach stand at a safe distance.

2.6. Outlook for the future: a EU-wide system?

Implementing a fully operational regional early warning system remains a very ambitious plan, and is beyond the scope of a research project like MICORE. The development of such a system from scratch would take several years and require end-user support at the national and European level. It has been found that often end users are generally not prepared to develop an early warning system on a regional scale even if they are interested in applications that demonstrate the capability of an operational tool. Thus, the operational prototypes developed and tested for each coastal site during MICORE were not maintained, with the exception of the Italian site in Emilia-Romagna, due to involvement of end-users since the very beginning of the project [HAR 13, HAR 16]. Furthermore, not all European countries already have an operational monitoring network for offshore oceanographic parameters, posing a barrier if they would like to integrate their data into an early warning system for coastal zones. In order to issue alerts relating to coastal hazards, early warning systems should be able to be based on real-time measurements of major forces (waves, water level, wind and currents) and characteristics specific to each coastal area (beach morphology, the presence of infrastructure, etc.).

Reducing the risks associated with coastal erosion and marine submersion involves an integrated management of vulnerable areas, and the development of more flexible defense structures. An appropriate disaster risk reduction (DRR) strategy also involves the implementation of adaptation measures (in relation to climate change and rising sea levels). Apart from structural measures, early warning systems also contribute, in the short term, to reducing the impact of storm surges through a coastal surveillance network and marine weather forecasts. In all cases, risk reduction measures related to extreme storms require an appropriate assessment of the degree of risk for a given coastal area. It is therefore important to define threshold values beyond which the risk can be considered significant enough to take the necessary measures to reduce the impact of storm surges. The impact indicators defined by the MICORE project made it possible to define threshold values based on physical (hydro- and morphodynamic) parameters, but did not focus on socio-economic parameters. The threshold values can be used in early warning systems to alert coastal populations to the potential dangers to which they are exposed and to prompt civil protection organizations to take the necessary measures to minimize the impact of storms. Some of these unresolved questions are currently under study within the RISC-KIT project [VAN 14],which aims to develop a series of standardized tools, including a Bayesian approach for identifying hot-spots [GUT 15].

Finally, a pan-European storm surge warning system is needed. Although limits on spatial and temporal resolution exist for such a large system, work done on climatic scenarios and surge behavior has proved to be promising [VOU 16]. Working at such a large scale, some compromise must be reached regarding a number of inputs to the model. First, the bathymetry to be used must be a global dataset (e.g. GEBCO), which can be eventually integrated with higher resolution European datasets like the “European

Atlas of the Seas". Second, it is difficult to determine the level of coastal protection, as there is no global dataset with the exact characteristics of all coastal structures. Third, if we want to estimate inundation, a global DTM like ASTER or SRTM must be used.

A EU wide system is currently being built within the ANYWHERE H2020 project (http://www.anywhere-h2020.eu) and involves cooperation between public and private research institutions. This system will become operational within the EFAS-European Flood Alert System [BAR 08, THI 09, ROO 11].

2.7. Conclusions

Recent years have brought to us many tools and techniques for assessing coastal risks related to storms. Observing technologies such as lasers or airborne radars (LIDAR, ENVISAT, RADARSAT, etc.) provide many imaging datasets that can be used to characterize the geomorphology of the beach, as well as the sea state. The recent launch of the EU Sentinel satellite programme has made free, high-resolution satellite products available to the wider community. Observing the geomorphological characteristics of the beach and dunes over time is essential for assessing the changes induced by extreme storms. Moreover, ground-based video surveillance systems allow regular monitoring of the evolution of the beach and more particularly of the changes in its sedimentary context and its hydro- and morphodynamic characteristics. Field data, which supplement the remotely sensed data, generally consist of DGPS surveys along transverse transects to better characterize beach profiles and possibly integrate them into coastal dynamics models. These should still be collected, for validation of remotely sensed products. Due to these *state-of-the-art* technologies, as well as new numerical models and formulas incorporating morphological characteristics of the beach and dunes, a better understanding of the influence of storm surges on coastal systems is possible. These considerations are essential if we are to assess as accurately as possible the current and future risks to which coastal populations are subjected. Historical data on past storms are also very important as it allows time series of weather data to be constructed for predicting the characteristics of future storms. Climate models using different scenarios for occurrence of extreme events and sea-level rise complement the forecasts and determine the real challenges for coastal risk management over the long term. In addition to physical factors, socio-economic and environmental factors should be an integral part of assessing coastal storm and flood risk. In this context, an accurate assessment of potential losses or real losses of goods and services (natural and human, tangible and intangible) following an extreme storm event is necessary to estimate the vulnerability of coastal dwellings. To this end, several methodologies for strategic risk assessments have been proposed. It should also be noted that an efficient DRR measure to reduce the impact of storms on coastal populations and the environment could only be based on a preliminary assessment of the risk of coastal erosion and marine submersion. The impact indicators developed by the MICORE and RISC-KIT project precisely measured this degree of risk. Coupled with marine weather forecasts, they enabled the development of Early Warning Systems applicable to different types of coastal sites. Finally, it should be noted that although research has made significant progress in this area, storm risk assessment remains a major challenge and a crucial step in risk management of coastal areas of the European Union and beyond.

3

Xynthia, February 2010: Autopsy of a Foreseeable Catastrophe

3.1. Introduction

In an exhaustive report entitled "Learning from French experiences with storm Xynthia, lessons for the Netherlands", published 6 months after the catastrophe by the *Ministerie van Verkeer en Waterstaat* (service of Dutch water), its director estimated without ambiguity that the crisis was badly managed by the local and national authorities, in particular because they did not regard the storms of the past as warnings. He stated thus that the catastrophic storm of 1953 which struck the Netherlands would from now on serve as a historical reference for the strategy of Dutch littoral defense. From this point of view, he thus considered it essential to study the French case in February 2010 because it was not dissimilar from Holland [KOL 10]. As he wrote, the purpose of his report was to give lessons for the Netherlands and not to give a perfect list of facts about the storm.

Six years after the catastrophe of February 27th and 28th 2010, this work aims to highlight the mechanisms of a natural disaster which had not been foreseen by local populations, their elected representatives and the French State. Among the causes explaining the high death and material toll of the Xynthia storm, the fact that they did not take into account the historical experience largely explains the increase in the exposure of littoral societies to these risks since the end of World War II. For the lack of integrated research associating the exact sciences (climatology, oceanography, modeling, etc.) and social sciences, in particular the history, in France, the results which follow proceed mainly from studies led within the framework of the historical report of inquiry (remained dead letter since) returned to the French Parliament on July 4th, 2010 and the European project FP 7 RISC-KIT (Impact strength-Increasing Strategies for Coastal-toolKIT) [GAR 10].

3.2. Scenario of the crisis

On February 28th 2010, at around 2 o'clock in the morning, Xynthia struck the French Western coast, mainly the portion of the Atlantic arc ranging between the south of Brittany and Bordeaux, causing the majority of the human and material losses in the French "départements" (or counties) of the Vendée and Charente-Maritime. In several places, dams, seawalls, dunes and other structures caused severe flooding on the Atlantic coastal. More than 50,000 hectares were flooded and the death toll rose to about 50 deaths.

3.2.1. *French coastlines*

The evening of February 23rd 2010, Météo France (French weather service) announced an active depression transforming itself into storm. Later on, this depression would be called Xynthia. At about midnight, be February 27th 2010, the storm hit the French coast with a Beaufort 10 gale (89–102 km/h). Beaufort 10 is not actually a very strong gale but the storm surge and the high tide coincided to cause high water levels and made it possible for the waves to cause a lot of damage. Indeed, usually wave action is limited to shallow areas off the coast. The high tide had a rating of 102 on a scale of 80 to 120, when the highest tides are normally expected in September. The return period of this flood was estimated to be of around 100 years, an estimation based on very short historical series.

The storm occurred in the south-west of Madeira, in the shape of a vast swirl accompanied by strong winds, rains and waves. On February 27th, the storm was reinforced and became an extratropical storm baptized Xynthia by meteorologists. Because of the decline of the atmospheric pressure, it became a phenomenon of explosive cyclogenesis also called "weather bomb" that intensified very quickly and moved faster than a hurricane [LIB 13]. It moved then from Portugal to Scandinavia, while crossing France around midnight on February 28th, according to a south-eastern/north-eastern axis (Figure 3.1). This storm hit the coasts of the Vendée and Charente-Maritime counties at the time of great equinoctial tides whose coefficients were higher than 100, the maximum being of 120.

Since its formation in the middle of the week, storm Xynthia was being observed thanks to satellites and weather forecasting models. Unfortunately, these did not make it possible to specify the intensity of the strongest winds, nor the progression of the storm in time and space before Friday, February 26th. On this day, Météo France extended a first message of alarm to the public and the media in the night. At 6 am on Saturday, the weather reports established by the national center of the Météo France forecast based in Toulouse and by the weather center of Rennes described an event of strong winds, a storm of force and exceptional width likely to disturb domestic activities and cause damage. However, weather forecasting was unable to estimate the behavior of the waves and the surge when they were to meet coastal infrastructures (dams, ports and dwellings). Worse still, no State services were called upon to verify the sea surge (Figure 3.2).

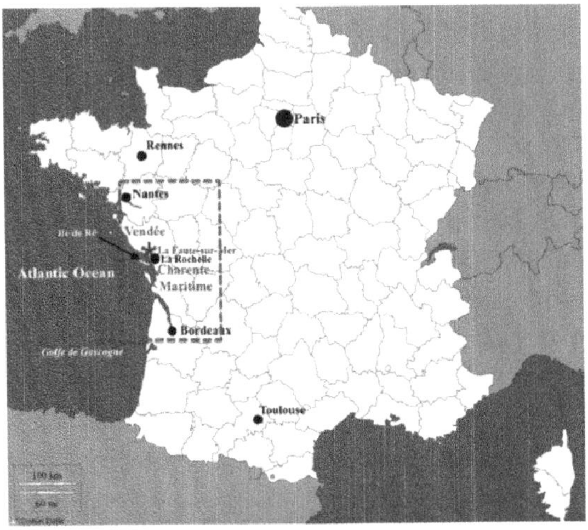

Figure 3.1. *Map of the Xynthia storm (February 2010). The green rectangle indicates the geographical zone affected by the storm. For a color version of this figure, see www.iste.co.uk/quevauviller/storms.zip*

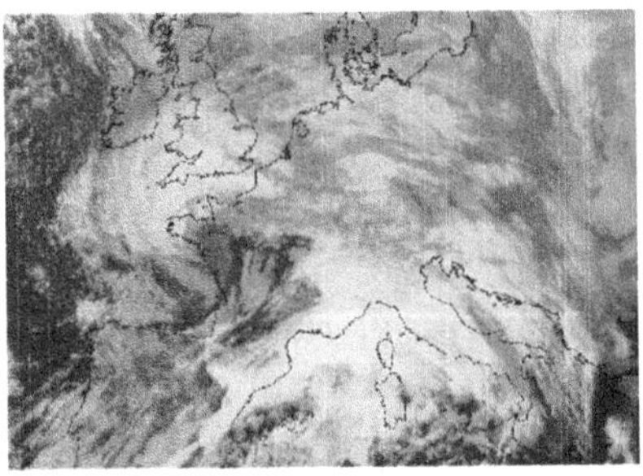

Figure 3.2. *Satellite image of February 28th, 2010 at 02:00 UTC. (source: Météo France[1]). For a color version of this figure, see www.iste.co.uk/quevauviller/storms.zip*

1 http://www.meteofrance.fr/climat-passe-et-futur/evenements-remarquables/la-tempete-xynthia-des-2728-fevrier-2010.

The relative intensity of the storm decreased gradually and moved toward the north of Germany and Scandinavia, disappearing completely on March 5th east of Saint-Petersbourg.

The singularity of Xynthia lies mainly in the impact of the storm surge on the sea level, with flooding which in particular affected the coasts of the Vendée and Charente-Maritime counties. The phenomenon of a sea surge can be defined as the flooding of areas by the sea that are normally dry. This risk of flood is induced by the forcing of several weather and maritime factors of origin which involve an abnormal rise in the marine level.

For Météo France, the phenomenon of marine flooding is also called "wash-over". It is dependent on four successive independent factors:

– a coefficient of strong tide generates a mean level of marine sea high called storm surge;

– then, a barometric depression plays a paramount role by giving rise to an atmospheric surge or storm surge. The latter is the combination of the barometric surge and the anometric surcote;

– with the approach of the littorals, transfers of wave energies generate a new rise in the water level [AND 13];

– finally, the action of the swash runup causes the surge of the waves on the littoral.

The risk of immersion also depends on the altimetry of the littoral zone affected by the waves. The lower altitude of the original ground, the greater the natural vulnerability. However, the most vulnerable sectors of the Atlantic coastline are located between the estuary of the Gironde and the estuary of the Loire, in particular the estuaries and marsh of the Charente-Maritime. These zones had the greatest flooding at the time of the passage of the Xynthia storm. The winds were higher than 110 km/h in Charente-Maritime, with gusts reaching more than 160 km/h locally.

However, a comparison with hurricanes Lothar and Martin in December 1999 and Klaus in January 2009 shows that Xynthia was definitely less intense. Consequently, its intensity does not come from the physical phenomenon but more from the unhappy conjunction of the winds and the high coefficients of tide on February 28th, 2010 which caused the immersion of the grounds. At the time, it seems that no model could envisage the sea surge which was of +1.53 m in La Rochelle and 1.75 m in the area of La Faute-sur-Mer [BER 12].

In terms of casualties, Xynthia was particularly heavy. In Europe, the storm killed 56 people and in France 47 people died. Xynthia is thus the most fatal catastrophe in France since 1992, date of the flood of Vaison-la-Romaine located in the Var county (42 victims). In addition, a strong correlation can be observed between mortality and housing since 37 deaths occurred in one-storey homes. The absence of a top floor was thus a factor of strong vulnerability. Another characteristic of mortality was the proximity to infrastructures, the totality of the deaths were located at less than 400 m from dams or seawalls strongly damaged by breaches. The study of the age of the victims also reveals that mortality mainly concerned individuals aged over 60 years, logically more vulnerable than of young individuals vis-a-vis the rise of water, and women. Lastly, five children died at the time of the storm, often victims of hypothermia [VIN 12].

The economic damage credited to the Xynthia storm is distributed between the damage caused by the winds (falling trees and roofs), the retreat of the coast line and finally the damage related to flood (dwellings, dams, etc.). The total loss, estimated by the Caisse Centrale de Réassurance (CCR:

French Reinsurance Company), was of about 1,500,000,000 euros, a lower cost, however, than that of hurricane Klaus in January 2009.

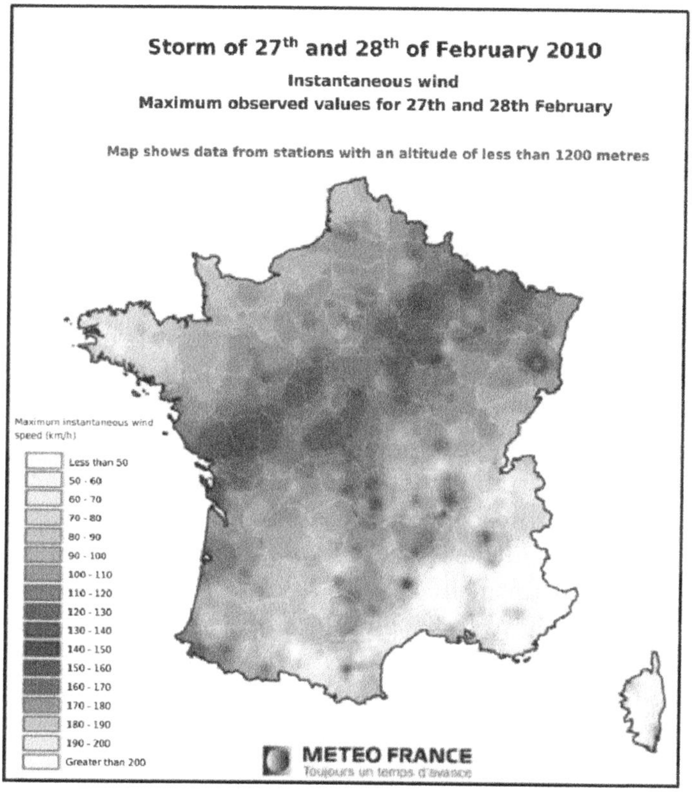

Figure 3.3. *Storm of the February 27th and 28th, 2010. Maximum wind speeds. (source: Météo France[2]). For a color version of this figure, see www.iste.co.uk/quevauviller/storms.zip*

3.2.2. *La Faute-sur-Mer: "martyred" city*

The seaside town of La Faute-sur-Mer, located in the county of Vendée, represents the national symbol of the

2 http://www.meteofrance.fr/climat-passe-et-futur/evenements-remarquables/la-tempete-xynthia-des-2728-fevrier-2010.

disastrous effects caused by storm Xynthia. The commune is located on a sandy peninsula of 500 m by 2 km broad. It is bordered in the west by the Atlantic Ocean and in the east by the estuary of the Lay river. The relief of this territory is thus low, most grounds having an altitude lower than 5 m NGF (general leveling of France).

Due to the minutes of the trial which resulted from the catastrophe and which were put online for educational purposes, it is possible to understand the chronological course of the catastrophe, hour by hour[3].

On site, the sea surge, or, in other words, the rise in the sea level during the storm, caused important breaches in the dune ridges and on the sea walls, in particular in the locality "Belle Henriette", north of the city. More importantly, because it was unexpected, the rise of the water was also related to the estuary of the river Lay, one of the largest of the Vendée. At the time of the storm of February 2010, the northern and southern sectors of La Faute-sur-Mer were protected from flood, in case of a large tide which would go up in the estuary, by a simple lifting of ground and fill, called the East Dike, built after 1850. However, at the time of the storm, the East Dike did not have the same height everywhere.

According to the surviving witnesses and the experts heard with the lawsuit, the first overflows of the East Dike took place from 3:00 am on Sunday, February 28 2010, a fact confirmed by the first call of a victim of the residential development of the Dory to the firemen at 3:24 am in order to warn them of the beginning of the flood.

According to expert testimonies, there were five locations of overflowing, that is to say a 580 m length, where the height of the dike lay between 4 and 4.20 m. No breach of the

3 http://www.jac-cerdacc.fr/xynthia-les-responsabilites-penales.

dike would occur this night. All the witnesses agreed that the water level rose very quickly, sometimes 1 m or more in only 10 or 15 min. These variations are explained by the very irregular topography of the site, with many basins which supported a strong progression of the flood in the streets. The overflow lasted 2 h with a peak of tide at 4:27 am. The surge reached 360 m a second, with a water wave of 80 cm at the lower point, and a rate of flow of 7 m per second[4].

The total volume of water contained in the southern basin of La Faute-sur-Mer was estimated at 1.2 million m^3. When the tide decreased, the water level in the low zones was stabilized at each rate of 3.80 ms NGF, and the evacuation was then very slow because the dike had become a trap by preventing the water run-off.

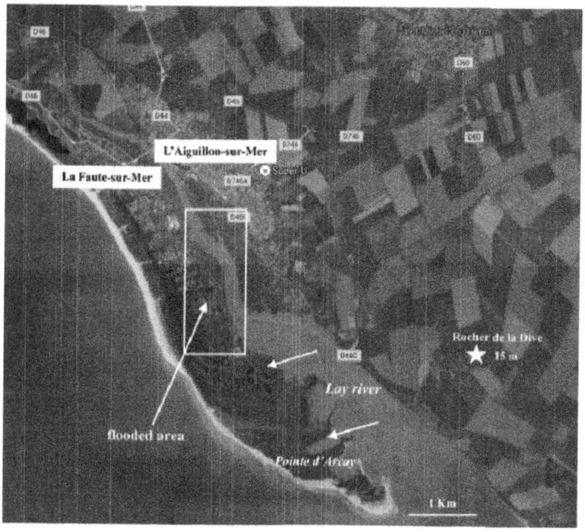

Figure 3.4. *View of the flooded zone (white rectangle) of La Faute-sur-Mer. The white arrows indicate the direction of the progression of water. The site of the Rocher de la Dive corresponds to a historical housing settled on a hill with 15 m above sea level. (source: Google-CNES/Spot Image and E. Garnier (comments)). For a color version of this figure, see www.iste.co.uk/quevauviller/storms.zip*

4 Records of the trial, p. 57.

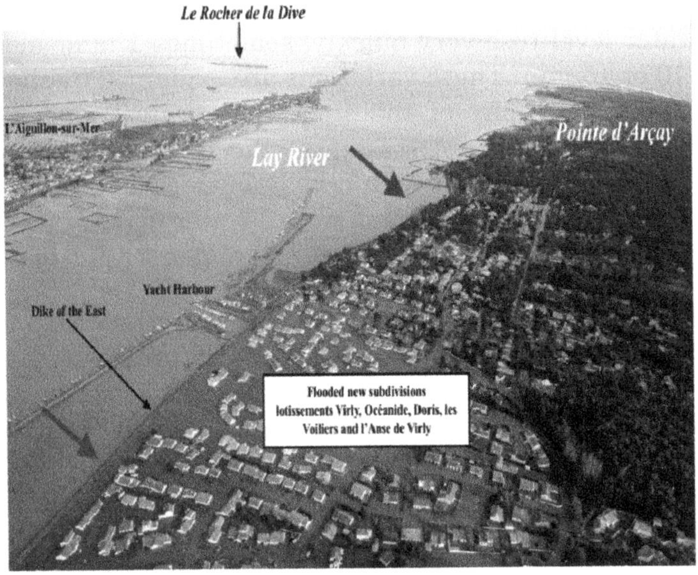

Figure 3.5. *Aerial photography of the flooded residential developments of La Faute-sur-Mer. The blue arrows indicate the direction of the flood since the Lay river. The site of the Rocher de la Dive becomes again an island like in the past. (source: Créocéan (photo) and E. Garnier (comments)). For a color version of this figure, see www.iste.co.uk/quevauviller/storms.zip*

This reality explains the high death toll. Twenty-nine people died that night:

– 10 men;

– 16 women;

– three children.

In addition to these deaths, there were 47 casualties and 33 hospitalizations. 767 people were also evacuated. Among these, 26 people resided in the very exposed residential developments. Three quarters of the victims were over 60 years old. The oldest was 87 years old. The forensic scientists who wrote the autopsies concluded that they had died by drowning, excluding four whose bodies were found days later near their homes. That night, the power of the sea burst bay windows, interior partitions and ceilings. This damage prevented many from escaping their houses or from closing doors and openings. Certain survivors told firemen that they had the impression of being in a washing machine. This description is corroborated by the multiple wounds observed on the corpses, which had received trauma before or after their death.

As regards the contingency plan, the firemen took action from the first distress calls, around 3:00–3:30 am. They came from L'Aiguillon-sur-Mer but were very few. They were already in a state of alert, but only for strong wind and not for sea surge. When they wanted to intervene, their barracks were flooded as well as the bridge which connected L'Aiguillon-sur-Mer and La Faute-sur-Mer. They could not help the victims quickly, in particular in the residential developments where they arrived at 6:30 am, despite the flood having started toward 3:30 am[5]. Finally, as they could not contact the mayor to organize help, they had to wait until the early morning of February 28th for reinforcements of the *Sécurité Civile*.

3.2.3. *The "unprecedented dogma"*

The press immediately seized the Xynthia catastrophe which became one of the first news stories to last at least a month. This can be understood by the extent of the losses and in particular the significant number of victims. But more

5 Record of the trial and testimonies of survivors for the historical report about the Xynthia Storm for the French Parliament and Senate Inquiry (July 4, 2010).

still, the exceptional mediatization of Xynthia was dependent on the atypical character of the catastrophe. The phenomenon of sea surge would be quickly regarded as exceptional by the journalists, as is frequent after each extreme event in France.

M Planète

Énergies Biodiversité Ressources naturelles Population Agriculture & Alimentation Polluti

"On ne peut pas prévoir l'imprévisible"

"Nous avons donné des consignes à la population et alerté les élus et les collectivités", raconte Béatrice Lagarde, sous-préfète des Sables-d'Olonne, en Vendée. La tempête Xynthia a fait au moins 48 morts en France (55 en incluant l'Allemagne et l'Espagne). Nicolas Sarkozy vient d'arriver en Charente.

Figure 3.6. *Extract of the national newspaper Le Monde of March 1st, 2010. Its title declared: "we can't predict the unpredictable"*

The political authorities and decision makers, as well as part of the scientific community, reacted with similar disbelief. The sous-préfet of the Vendée stated that "we cannot envisage the unforeseeable" only 3 days after the disaster. A minister declared for his part by March 1st on National radio RTL that the coincidence of the storm and the tides was unforeseeable, concluding that "it was a phenomenon which we have not seen for several centuries". For its part, the French Geological Surveys (BRGM) estimated that the probability of occurrence of such an event would be approximately every 10,000 years. Later on, the engineers of the BRGM specified their declaration by explaining that their statistics were obtained starting from series collected on a few tens of years. In addition, the certainty it was about a new event was reinforced by the recent publication of a report announcing a rise in 1 m of the sea level for the Atlantic coast.

However, neither the media nor the politicians remembered the great storm of 1953 which struck the European countries of the North Sea and which caused thousands of deaths. In France, people and policy makers completely forgot that this storm had affected the littoral of the French coasts between Dunkirk and Calais, in particular in Gravelines where a nuclear power plant was built in the 1980s, despite the fact that the local press of the time testified to this extreme event precisely.

DUNKERQUE

Au cours de la nuit de samedi à dimanche,

LA TEMPÊTE A CAUSÉ D'ÉNORMES DÉGATS DANS TOUTE L'AGGLOMÉRATION

La machinerie de l'écluse Watier a été noyée

La côte qui, à la digue Tixier, devait être de 5 m. 80 s'est élevée à 8 mètres

MARINE A DUNKERQUE

Figure 3.7. *Extract of the newspaper "La Voix du Nord" of February 4th, 1953 describing the damage of the storm in Dunkirk and Gravelines*

Yet, more than fifty years later, the mayor of the city of La Faute-sur-Mer affirmed again, during his lawsuit, that the deaths of the victims were the consequence of an "unforeseeable and exceptional" event. Sentenced for the first time to 4 years of imprisonment on December 12th 2014, he was finally sentenced on appeal to 2 years on a conditional sentence on April 4th 2016.

Like the mayor, this mediatic and political position on the concept of exception were already used in December 1999 after the Lothar-Martin hurricane and in the hours which followed the catastrophic floods of Draguignan in June 2010 and Mandelieu-la-Napoule in October 2015.

3.3. The historical verdict

Vis-a-vis the systematic assertion that the Xynthia catastrophe was unprecedented and unforeseeable, it is time to turn to the history. Indeed, like European project FP 7 RISC-KIT showed, history offers an important documentation likely to reveal the chronological reality of this kind of extreme event from as early as the 16th Century, in France as in the rest of Europe.

3.3.1. *At the national and European levels*

The historical study of sea surges on the French coastline between 1500 and 2010 that we present in this section does not aim to be exhaustive. So, certain catastrophes may not be listed, for lack of information in the archives. However, those collected to date are certain. Consequently, the results suggested in terms of frequency return period are a low-end estimation but confirmed in the documentation.

The study discredits the idea according to which the storm of February 2010 was a completely unforeseeable risk. Indeed, between 1550 and 2010, 117 sea floods were documented in the archives for the whole of the French coastline. Among them, 30 catastrophes struck the Atlantic coast exclusively. More interesting are the periods of return calculated from these series. Once again, they are relatively homogeneous since they are included between 14 and 19 years, with a risk estimated at 15 years on the Atlantic shores, whereas the French Naval Hydrographic and Oceanographical Service (SHOM) and the Radioprotection and Nuclear Safety Institute (IRSN) propose return periods from 50 to 100 years [PIN 10, BAR 11].

Littoral	Number of occurrences (1550–2010)	Return period (years)
Atlantic Coast	30	15
Channel	29	16
Mediterranen Sea	24	19
North Sea	34	14

Table 3.1. *Historical sea surges of the French coast (1550–2010)*

Focused on the Atlantic coastline, the chronology shows strong disparities in terms of floods in the last 500 years. Contrary to what might appear to be the case, the last 50 years did not know recrudescence of this kind of extreme events. The most catastrophic century corresponds to the 18th Century with nine sea surges, whereas the 20th Century had only five. It also seems that the apogee of the Little Ice Age of the 17th Century resulted in a lower frequency of these extreme events (three). During the last 100 years, six sea surges struck the French regions of the Atlantic with a notable characteristic nevertheless: the majority occurred between 1924 and 1957.

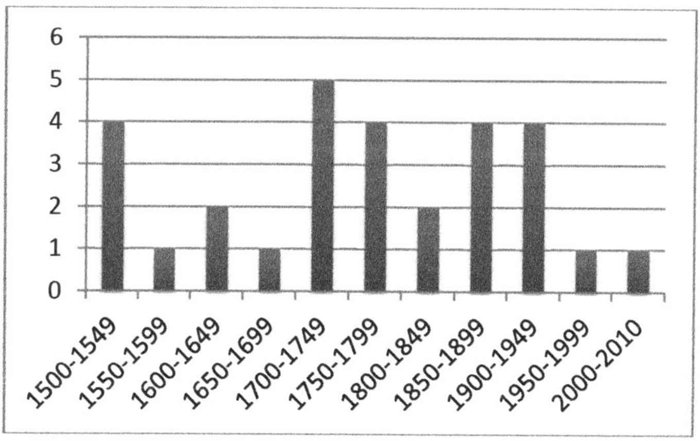

Figure 3.8. *Number of sea surges per 49-year periods on the French Atlantic coast (1500–2000). For a color version of this figure, see www.iste.co.uk/quevauviller/storms.zip*

On March 15th 1937 the newspaper *L'ouest-Eclair* ran the headline: "true tidal wave breaks on the Atlantic coast". In fact, it was an exceptional weather event (much like the Xynthia storm) whose force appeared unusual ("exceptional" as said in the press) insofar as the scenario of the catastrophe involved a very violent storm of about 973 hPa (975 hPa in 2010) with strong coefficients of tide of 109.5 in Brittany and 102 in the port of La Rochelle (102 in 2010). The storm hit the Atlantic coast on a south-south-west axis (identical to that of Xynthia) and caused significant damage at La Faute-sur-Mer (sea walls cut in several places), but without human loss because of a better coastal land settlement.

UNE TEMPÊTE D'EQUINOXE D'UNE RARE VIOLENCE

Les flots en furie ont déferlé sur les côtes de l'Atlantique causant d'énormes ravages

SUR LA COTE VENDÉENNE, C'EST UNE TERRIBLE CATASTROPHE : LA LOIRE-INFÉRIEURE A ÉTÉ TRÈS DUREMENT ÉPROUVÉE

De nombreuses digues ont été rompues ; des centaines de personnes ont dû évacuer leurs habitations et les dégâts atteignent plusieurs millions

La presqu'île guérandaise et la région nazairienne sont dévastées et les dégâts qui ne peuvent être encore évalués, sont considérables

Figure 3.9. *Extract of the local newspaper L'ouest-Eclair of March 15th, 1937 with the headline: "furious floods broke on the Atlantic coasts causing enormous devastations… on the Vendean coast, it is a terrible catastrophe"*

After this, no event of similar intensity struck the coast or its population. Despite being very powerful, the Lothar and

Martin storms of December 1999 did not cause significant flooding, except in the sector of the nuclear power plant of Blaye (near Bordeaux). For this reason, the authorities may not have recalled this event at the time of Xynthia [GAR 11].

Thus, between 1957 and 2010 (Xynthia), in other words for more than 50 years, populations on the Atlantic coast, their elected representatives and the French State did not have to cope with the risk of flooding, contrary to Northern Europe. Indeed, the northern neighbors of France had to deal with storms in 1953 and 1962 [LAM 91]. The catastrophe of 1953 caused the death of 2,100 people including 1,800 only in the Netherlands. The event took place the night of February 1st, when a storm of the North-West affected these northern countries. The combination of a strong coefficient of tide and wind contributed to push water at the point of reaching 4.5 m above sea level. In Holland, the sea walls were broken in many places along on the west coast, as such the sea invaded very wide territories. The houses close to the sea walls were completely destroyed by the waves and their owners completely surprised by the flood in their sleep. This explains the high death toll to which are added 72,000 evacuated people, very damaged road and communications networks because of the fall of telephone posts. Everywhere, the sea forced firemen and survivors to move with boats [GAR 15b].

For the east coast of England, the agency of the British environment estimated that 300 individuals perished, 24,000 houses were destroyed and 40,000 people were evacuated. In Belgium, many sea walls were flooded or broken, causing the flooding of Ostend and Antwerp and the death of 40 people.

Year	Month
1924	January
1937	March
1940	November
1941	February
1957	February
2010	February

Table 3.2. *Sea surges of the Atlantic coastline 1900–2010*

The wave which flooded Hamburg during the night of February 16th 1962, like Xynthia in February 2010, came after a long period of little littoral risk since the last known storm of this kind was in 1855. However, during this long period, the defense systems of the sea walls and dikes worsened considerably because of a lack of maintenance due to a lack of financial considerations and a misleading feeling of invulnerability. That is why the flood there was particularly violent, with more than 300 lives lost in Hamburg because of the lack of this risk for approximately a century [STO 06].

Unfortunately, France did not learn any lessons from the experiences of its neighbors. Worse still, the memory of these tragic dates seems to have faded, a process facilitated in France by the "catastrophic remission" (absence of sea surges) after the 1960s. This may partially explain the general feeling that Xynthia was unforeseeable.

3.3.2. *The example of La Faute-sur-Mer*

As previously mentioned, the territory of La Faute-sur-Mer is naturally exposed to sea surges risks because of its low altimetry and situation in an estuary.

The historical documentation of storm surges in La Faute-sur-Mer is extremely difficult because this coastal city did not exist on the administrative level before the second half of the 20th Century. Indeed, La Faute-sur-Mer became a municipality only in 1953. The populating of the site only began in the 19th Century under the shape of some fishermen's houses forming a hamlet [GAR 15b]. As a result, it seems unrealistic to propose a definitive chronology of storms for 17th and 18th Centuries. Certainly, archives evoke disasters before 1800, but it is impossible to affirm that they directly affected this case study.

It is thus necessary to discriminate the certain events, for which we have precise historical data directly concerning the site, from the uncertain events. The latter indicate storms that had a regional impact (counties of Charente-Maritime and Vendée) without the guarantee that La Faute-sur-Mer was concerned.

As for the whole of the Atlantic coast, the chronology of the sea surges of La Faute-sur-Mer is characterized by a strong temporal variability. The historical approach carried out for European project RISC-KIT allowed us to collect with certainty seven examples of floods by the sea at the time of storms (Review report RISC-KIT Project, 2015).

December 9th 1711, during almost 9 h, the Vendée and Charente-Maritime regions were struck by a powerful storm surge which corresponded to a tidal range of 97 (La Rochelle). The sea level reached 6.34 m in the morning. Considerable damage was listed in the region. Archives evoke the destruction of numerous seawalls, the flooding of saltworks, which are vital for local prosperity. The authorities estimated the cost of the disaster at 1,185,344 € because it was necessary to reconstruct seawalls for years. If the event was certain on a regional scale on the other hand, no historical source allowed us to affirm that it struck the site of La Faute-sur-Mer which was totally uninhabited in 1711. Archives talked especially about La Rochelle and the Île de Ré which, were severely affected.

The storm of February 21st 1788 was probably one of the most violent storms of the maritime history of the region (Charente-Maritime and Vendée). Caused by a South-West wind during the full moon, the storm submerged the coast in several places. The barometric pressure observed in La Rochelle was estimated at 9,764 hPa in the afternoon. In this city, several districts close to the port were flooded by the sea. The floods also affected islands (Oléron, Ré) and the city of Les Sables d'Olonne. Almost everywhere, seawalls were broken. It is likely that this surge storm affected the current sector of La Faute-sur-Mer unfortunately, no historical proof can prove it because the site was totally deserted.

The surge storm of January 1st 1877 was associated with a tidal range of 94 on December 31st in the evening. It was noted in the registers of the Conseil Général de Vendée and in the archives of the city of l'Aiguillon-sur-Mer. It most probably affected La Faute-sur-Mer. Again, historical proof is lacking for La Faute and archives describe only the damages caused to the municipality of La Tranche-sur-Mer. The map of dunes drawn in 1878 by the Forestry commission shows for the first time a population of about 10 dwellings, essentially fishermen's huts.

The first storm surge mentioned in the archives in La Faute-sur-Mer occurred on October 27th 1882. During a high tidal range (110 in the evening), the swamps are submerged by the sea from the Pointe d'Arçay to the hamlet of La Faute and the Port Puaut. In terms of damages, archives evoke 444 ha flooded and 40 m of cut sea wall.

L'Ouest-Eclair

JOURNAL RÉPUBLICAIN DU MATIN

25 MARS 1928

25 CENTIMES

DE QUELQUES RESPONSABILITÉS DU CARTEL

UNE PARTIE DU LITTORAL VENDÉEN MENACÉ PAR LA MER

Plus de 120 hectares ont déjà été gagnés par les eaux. Mais le village de La Faute, au sujet duquel on avait de sérieuses craintes, paraît, momentanément, grâce aux travaux de fortune exécutés, hors de danger.

22 wagons d'explosifs arrêtés à la frontière yougoslave

Ces wagons venaient d'Italie

Figure 3.10. *Press article of March 25th, 1928 showing the damage caused by an immersion on the current territory of La Faute-sur-Mer. At that time, the sea invades only agricultural lands because of a weak urbanization*

As the press of the time demonstrated, the storm surge of March 23th–24th 1928 corresponded to a major disaster for the village of La Faute-sur-Mer, as proved it by moreover the media coverage of time. The instrumental data available are numerous. The wind was directed southwest and the tidal range estimated to 110. Collapses of sand occurred on the left bank (L'Aiguillon-sur-Mer) of the river Lay. Sources do not mention the evacuation of the population, but the village of La Faute-sur-Mer would from now be on an island. The threat was so grave that more than 450 inhabitants of the nearby municipalities (l'Aiguillon, Grues, Saint-Denis) were mobilized to supervise the seawall of Fenouillet which protected 7,000 ha of farmlands.

Archives and press clippings show that the sea dug a channel in the dune which created a first breach of 200 m by which waters invaded the lands. The second breach was opened and waters flowed to the farm of "La Violette". The result was disastrous. The sea flooded more than 120 ha with farmlands and cut the road between La Faute and La Tranche-sur-Mer for or 1.5 km. Dunes collapsed and almost disappeared, whereas pines planted by the forestry commission were uprooted and taken by waves.

A second event in La Faute-sur-mer, the storm surge of March 13th 1937, caused only limited damage. On the other hand, the rest of the French Atlantic coast (from Brittany to the Spanish border) was severely hit [GAR 15b]. On the meteorological level, the event corresponded to a violent storm due to a strong equinoctial tide wave. In La Faute, a seawall was broken (20 m breach) but quickly repaired due to the help of soldiers and inhabitants. These quick repairs limited the flood to the low fields situated behind the seawall.

Only a few years later, La Faute-sur-Mer was again the victim of a storm on February 16th 1941. South-West winds and a tidal range of 98 flooded the polders of the right bank of the river Lay, upstream of the l'Aiguillon bridge. The road between La Faute and La Tranche-sur-Mer was submerged by 0.60 m of water and the river Lay overflowed. A strong erosion to the right seawall of the Lay river was also observed. It was necessary to remove the sand that had accumulated against the dike up to a height of 2 m.

Figure 3.11. *Photography of the newspaper L'Ouest-Eclair showing the breach in the dunes of La Faute-sur-Mer (L'Ouest-Eclair Newspaper, 25 March 1928). For a color version of this figure, see www.iste.co.uk/quevauviller/storms.zip*

Figure 3.12. *Photography of the newspaper L'Ouest-Eclair showing the flooded fields and the road cut between La Faute and La Tranche-sur-Mer (L'Ouest-Eclair Newspaper, 25 March 1928). For a color version of this figure, see www.iste.co.uk/quevauviller/storms.zip*

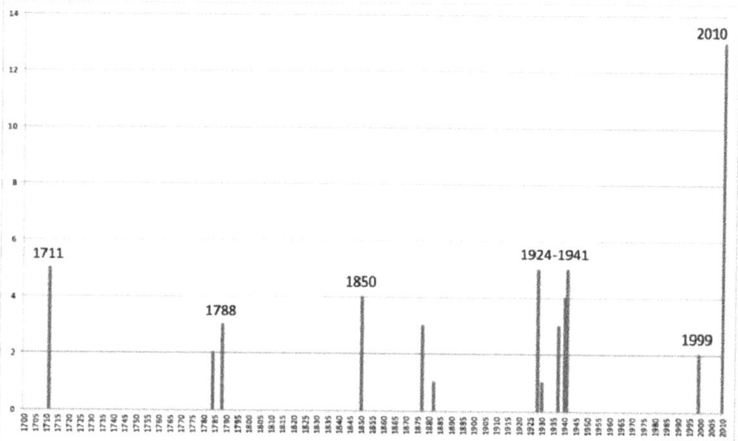

Figure 3.13. *Sea surges of La Faute-sur-Mer 1700–2010. For a color version of this figure, see www.iste.co.uk/quevauviller/storms.zip*

The history of the extreme events of the locality thus puts a definitive end to the debate, which took place after the catastrophe of 2010, about the rarity of this kind of extreme event. At least seven sea surges have affected the commune since 1711, that is to say a return period of 42 years. Today, the modelers and engineers estimate for their part that the local risk is of every 60 years. More astonishing is the fact that the commune of La Faute-sur-Mer suffered three floods in the first half of the 20th Century (1928, 1937 and 1941), a rather recent chronology which seems, once more, to be completely forgotten at the time of the tourist development of the city in the 1980s.

3.4. The construction of the coastal vulnerability

"Most of our physical ills are our own work"

This prestigious quotation of the French philosopher Rousseau concerning the earthquake of Lisbon in 1755 summarizes the challenges of the coastal vulnerability perfectly. In connection with the tsunami which flooded the city, he denounces for the first time the vulnerability of a coastal town. He then proposes a choice extremely radical but still current: to limit the urban concentration on the littorals by calling upon the precautionary principle.

3.4.1. *The time of the precautionary principle (Middle Age – 1900)*

For the continental populations and elites, coastal areas and the ocean were often perceived as hostile. Their inhabitants had a very bad reputation. They were thought to be "wreckers", i.e. people who mislead the ships by lighting fires on the beaches to cause their shipwreck and then to plunder them after having massacred their crews.

The royal State was interested for its part in the coast only for strategic reasons because the French coasts were exposed to the risk of invasion of its Spanish and British neighbors until the 1850s. That is why it was regarded as a space of uncontrolled borders because it was wild on the landscape level. Nevertheless, as of the 17th Century, because of the new naval ambitions of French monarchy, coastal space was punctually militarized with fortifications and military harbors, like those of Brest, Rochefort and Lorient, in order to welcome the royal fleet.

For coastal populations, the perception of their environment was completely different. The sea side is first and foremost a nourishing space exploited in a sustainable way, according to a requirement impossible to circumvent: safety. The foreshore is an extension of the villagers' landholdings and the conflict of uses between the communities are frequent. Indeed, the stakes are high in this zone of contact between the land and the sea. The survival of the inhabitants is largely conditioned by the fishing season, practiced directly on the shore (fisheries). These infrastructures are built with stones taken from the rocky outcrop, making it possible to capture fishes when the ocean is at low tide.

Figure 3.14. *Fishery of Le Phare des Baleine (Ile de Ré) in 2015 (source : E. Garnier). For a color version of this figure, see www.iste.co.uk/quevauviller/storms.zip*

The economy of the Atlantic coast was also based on grains, vineyards, salt made in vast salt marshes, the harvesting of wrack, the dredging of oysters and mussels (before the modern shellfish farming) and the collection of sand and pebbles for construction.

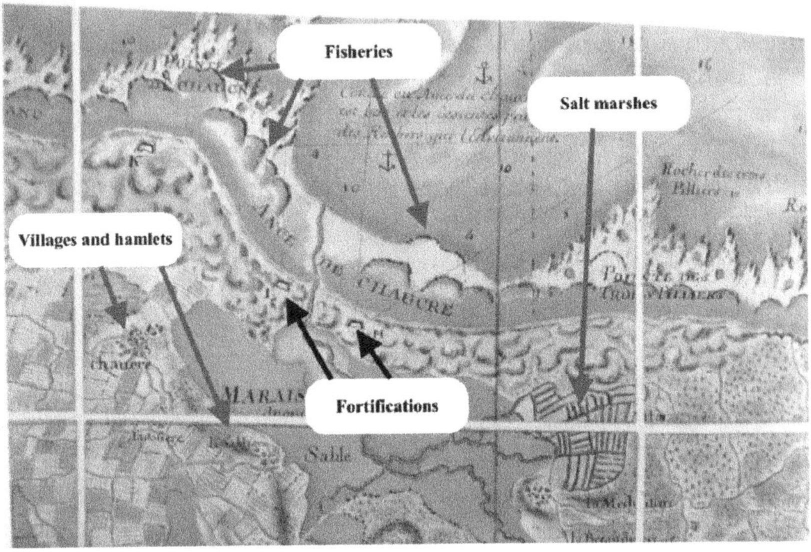

Figure 3.15. *Old map of the Charente-Maritime coastline (1757). Médiathèque de La Rochelle, 1 PL 128 (source : E. Garnier). For a color version of this figure, see www.iste.co.uk/quevauviller/storms.zip*

The militarization of the coastline by the monarchy resulted in an early map which makes it possible to have a very precise idea of these old littoral landscapes. As shown in Figure 3.15, the littoral dunes are represented perfectly in the shape of massive dunes which frame a marsh. The only visible traces of human impact on this dune are the forts and the multiple fisheries of semi-circular form. In addition to their already mentioned nutrional function, they protect the dunes by playing the role of obstacles against the swell. Some fisheries are still used on the Ile de Ré. Others were destroyed by the State because they obstructed navigation after 1850.

Finally, the villages and hamlets eventually settled on lands located above sea level. The wetlands also take part in the defense of the coast line by serving as protective zones against river floods and sea surges [GAR 15b].

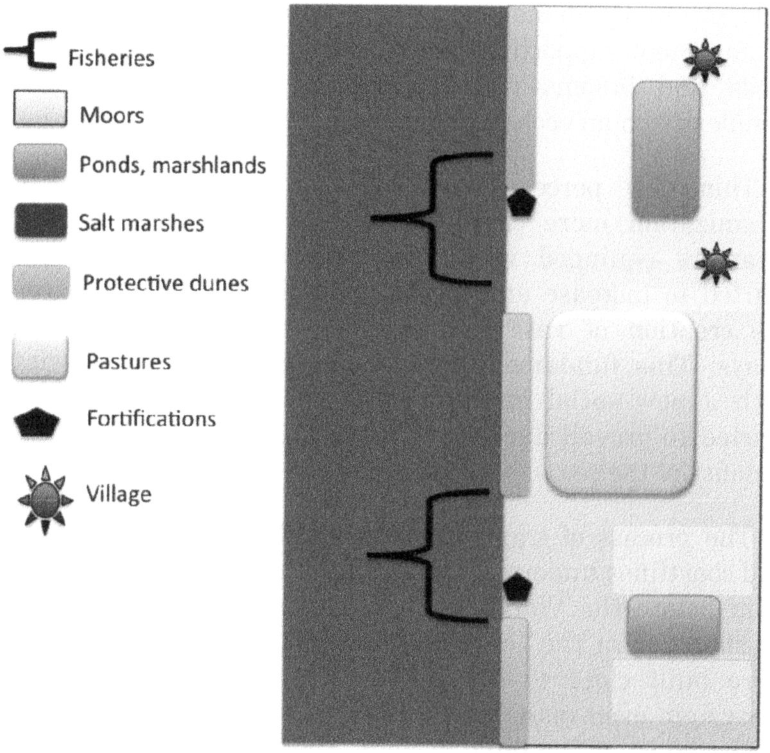

Figure 3.16. *Atlantic coastal management 1500–1900. For a color version of this figure, see www.iste.co.uk/quevauviller/storms.zip*

3.4.2. *The choice to live close to the sea (1900–2016)*

From the 1850s, most of the migratory dunes, in particular those of the Golfe de Gascogne (Bay of Biscay), were subject to massive afforestations of maritime pine. According to the forestry experts, it served to "domesticate" the littoral dunes. Gigantic work begun to build motionless and protective dunes. This policy of installation started again after World War II, this time in the form of a "mechanical replanning" and it was necessary to wait until the end of 1970 for this dune enterprise to be gradually abandoned so as to profit from a more durable management supporting semi-natural dunes (plantation of sea reeds and brooms with brushes and heathers) of greater complexity on an ecological level.

This new perception of coast lines supported a new urbanization more close to the sea. With the rise of the steamers equipped with iron hulls, the number of ports started to increase and modify strongly the coast because of the creation of vast basins, quay levels, warehouses and quays. This fundamental economic transfer was associated with a new social interest for the coast. A more ludic use started to prevail due to publicity in favor of the therapeutic benefits of the sea and as a the passion of artists and writers.

The process of transformation accelerated from the 1960s and coastlines underwent modifications much more disastrous than before the World War II, in particular because of the construction of the German Atlantic wall. New coastal roads were built close to the shore. The railroad disrupted the ecosystem even more because it was built closest to the shore in the heart of the dune system. The railway thus became a barrier cutting the dunes and worsened the impact of the floods by creating a basin effect and by disturbing the hydrographic network. These new transport routes served from now on urbanization poles made up of villas and allotments by the seaside. They were built directly in the dunes which had remained uninhabited up to that point, depopulated under the terms of an ancestral precautionary principle. From now on, these new inhabitants completely naïve about littoral risks were joined in summer by many tourists who required a direct access to the beaches. Everywhere in the dunes were open passages, while the trampling of the visitors spreads.

Figure 3.17. *Photography of the beach of Royan (Charente-Maritime) around 1920 (source : Archives départementales de Charente-Maritime, 24 Fi Royan 3). For a color version of this figure, see www.iste.co.uk/quevauviller/storms.zip*

Figure 3.18. *Photography of the beach Châtelaillon (Charente-Maritime) around 1900 (source: Archives départementales de Charente-Maritime, 10 Fi 123, n° 23)*

All these factors then joined together for an accelerated degradation, often perceived wrongly as an effect of climate change while at the same time it proceeds rather of the speculative development of last 40 years. This inexorable process occurring since the 1960s must be seen as the roots of the contemporary coastal vulnerability. On the matter of degradation, a well-known academic case of this exposure to risks is without question the city La Faute-sur-Mer.

On the Channel coastline (Known as *la Manche* in French), an increase in the urbanization by about 11% between 1990 and 2010 was currently observed as compared with 10% for the Atlantic coast and 13% for the Mediterranean shore. That is to say, an evolution of about 11% on the whole of the French coastlines. Between 1990 and 1999, this artificialization slowed down somewhat. Since 2000, increased land pressure accelerated very strongly the rate of artificialization.

Currently, a level of saturation in terms of the urbanization of the littoral communes has been reached. By way of an example, 98% of the littoral of the county Alpes-Maritimes (Provence) is urbanized.

Today, a little less than 700 communes on the whole of the French coast present a risk of marine flooding. The *départements* with the highest degree of exposure are those in dark blue in which we find the two which were the most impacted by the Xynthia storm (Vendée and Charente-Maritime). In terms of the Mediterranean shore, 60% of the communes present a risk of marine flooding, 63% on the Atlantic coast against 40% on the Channel. In total, in France, one coastal commune in every two presents a risk of sea surge with variable degrees of exposure.

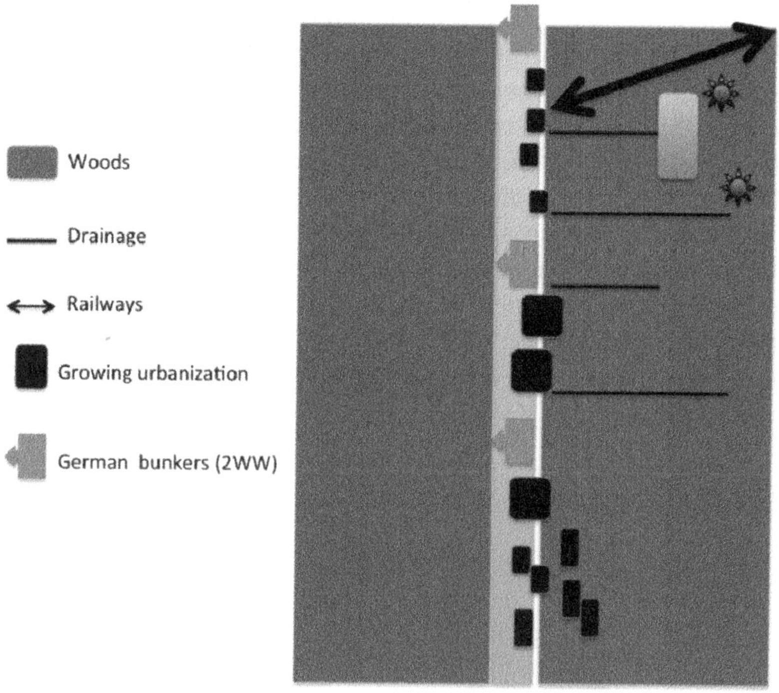

Figure 3.19. *Atlantic coastal management 1900–2016. For a color version of this figure, see www.iste.co.uk/quevauviller/storms.zip*

By comparing the degree of exposure of the French coastal communes to the lastest data of the French National Institute of Statistics and Economic Studies (INSEE) of population and residences, an estimate shows that more than 560,000 people live in these high-risk zones, that is to say an increase of 10% of exposed individuals since 1990. The number of exposed houses is estimated at more than 200,000, that is to say +20%, mainly attributable to the significant development of holiday homes [GAR 12].

By studying the historical storms and the values of exposure of reinsurers, we can confirm the identification of two particularly exposed regions. First is, without surprise, the Vendean coast with the cities of La Faute-sur-Mer, L'Aiguillon-sur-Mer, etc. The second corresponds to the Languedocian coast, subjected to a strong demographic pressure. By studying the zone of impact subjected to storms which caused marine flooding, a coastline particularly at risk emerges: Sète to Agde, located primarily on sandy low coasts very largely altered to the detriment of the dunes and ponds, and thus abound with marine intrusions. In this statistically defined perimeter, the most exposed cities are the cities of Frontignan, Agde, Marseillan and Sète with an estimate of more than 20,000 exposed houses.

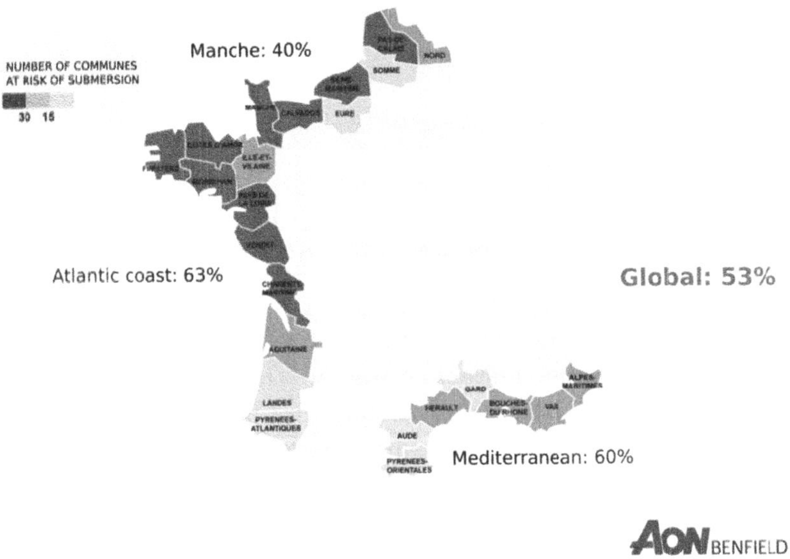

Figure 3.20. *Communes exposed to the risk of sea surge (source: Garnier et al., 2012). For a color version of this figure, see www.iste.co.uk/quevauviller/storms.zip*

3.4.3. *A national symbol: La Faute-sur-Mer*

Emblematic to this process of urbanization and its corollary, the artificialization of the coast, is the region of La Faute-sur-Mer and L'Aiguillon-sur-Mer. The current site of La Faute corresponds to a narrow sandy territory created in the 17th Century by a massive sea sand contribution supported by the context of the little Ice Age. Around 1760, the site was still covered by the river. The city nearest to the sea at that time was St-Michel-en-l'Herm which was protected by the distance and the marshes.

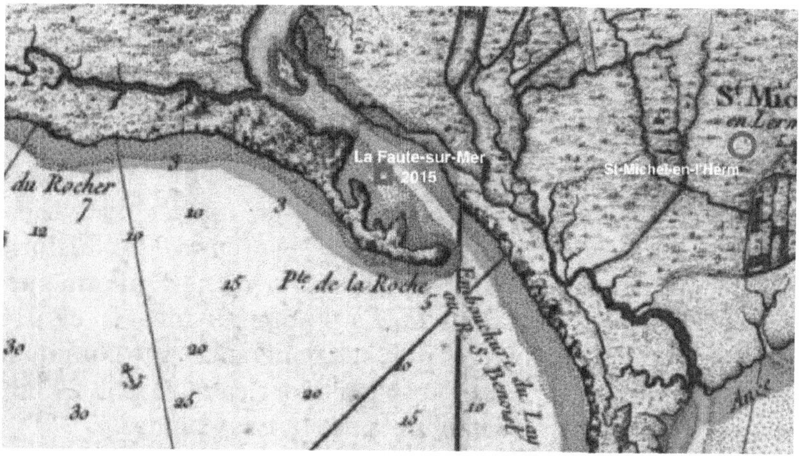

Figure 3.21. *Old map of the Vendée coastline drawn up in 1768. One can observe the formation process of the Pointe d'Arçay in a finger structure. The current site of La Faute-sur-Mer indicated by a red square is still located under water of the river Lay. Departmental records from the Vendée, 24 Fi. For a color version of this figure, see www.iste.co.uk/quevauviller/storms.zip*

In 1829, a hamlet was created by some farmers and in 1889, La Faute-sur-Mer did not yet exist as a municipality. The hamlet was made up of approximately 150 inhabitants and it was only in 1953 that the commune, due to part of the territory of the commune of La Tranche-sur-Mer, was created. The only locality of importance was the city of L'Aiguillon-sur-Mer, prudently cut off behind the strip of land which became La Faute and the dune ridge bordering the east shore of Lay river.

After World War II, the villas and the installation of the beaches started in order to accommodate the tourists coming from Brittany and Paris. However, this first tourist phase of installation preserved the natural zones. The aerial photographs thus showed very few houses, dispersed among agricultural lands. The coastal system was not yet degraded, with thick and high dunes, even if some of them are already cut in order to create direct passages toward the sea. In the Summer, the beach was occupied by many beach boxes which would be dismounted for the winter in case of storms.

Things really changed from the 1970s. The territory was gradually urbanized with the rise of tourist activities. Everywhere, constructions multiplied around the historic village, in particular with the creation of new allotments. Nowadays, the population of La Faute-sur-Mer is of 916 inhabitants, 46% of which are pensioners. In the Summer, the population of La Faute-sur-Mer can reach 20,000 inhabitants where 98% of houses are holiday homes.

The Xynthia storm made it possible to go ahead with some aberrations as regards town planning. To quote some figures, in La Faute-sur-Mer, 3,000 houses were built on old marshy zones including one major part which was under sea level. While disguising the risks involved, the developers made these sectors less resilient because they did not take into consideration the dangers of sea surge and erosion.

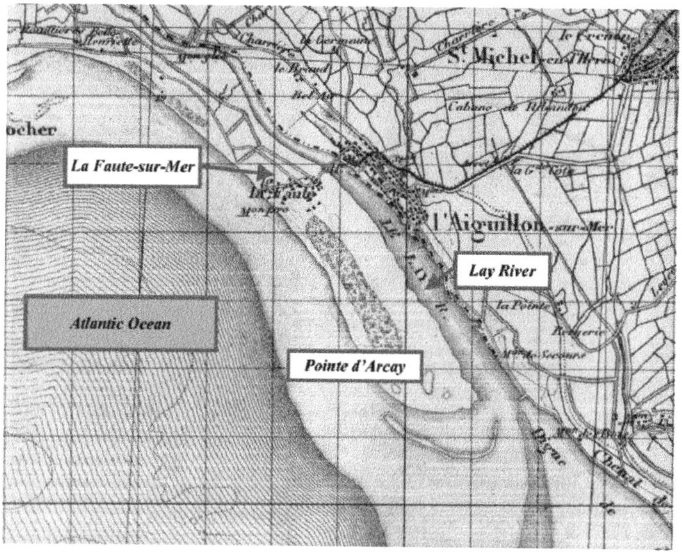

Figure 3.22. *Map of La Faute-sur-Mer in 1880. About 10 houses and farms protected by the natural dunes exist at that time (IGN 5EM141SO). For a color version of this figure, see www.iste.co.uk/quevauviller/storms.zip*

Figure 3.23. *Aerial view of the village of La Faute-sur-Mer in 1948. The habitat remains scarce and is composed of farms and some houses rented by tourists. The majority of the offshore area corresponds to agricultural lands, while the dunes are still well preserved (Archives départementales de Vendée, 2Num268). For a color version of this figure, see www.iste.co.uk/ quevauviller/ storms.zip*

Figure 3.24. *Urbanization of La Faute-sur-Mer in 2015 (source: Google-CNES/Spot Image and E. Garnier (comments)). For a color version of this figure, see www.iste.co.uk/quevauviller/storms.zip*

At the time of the Xynthia storm, the East dike built in the 19th Century did not have the same height everywhere. Its profile longitudinally was actually notched, like expert testimonies in 2006 and 2008 proved. More serious for the future, these same experts explained that the dike had been built on sand and mud. In fact, the zone where the allotments were located corresponded to a drained marsh. In the event of a sea surge, the risk of breach was obvious. The use of the word "dike" was very exaggerated because actually, it was rather a slope or earthy wall, terms used by the inhabitants themselves.

It was precisely behind this dike that were located the allotments which were the worst hit during the storm, i.e. the private housing developments of Virly, l'Océanide, les Doris, les Voiliers and l'Anse de Virly. In a surprising way, the top of the dike here was between 4 and 4.20 m. Consequently, it was the lowest part of the dike which was suppoed to protect the new dwellings. The subdivisions Dory and Les Voiliers were the two most recent real estate transactions of the southern part of the commune, during the first decade of the 2000s. The original ground of these two subdivisions, after its fill, was located at the NGF level of 1.80–1.90 m, i.e. below the level of the other side of the dike. This part was located at the NGF level of 2.60 m. Incredible but true, these contemporary subdivisions were built in a zone which had been the major bed of the river Lay. More incredible still, in 2002, a research department specialized in maritime engineering was charged by the State with the development of a Flood Prevention Plan (PPRI) in this area. It proved that this part of the city was a large basin where water would accumulate very quickly in case of the rupture of the dike or an "overflow", i.e. overflowing of the dike. It was exactly this second phenomenon which occurred on the night of February 27th and 28th 2010[6].

This chapter aimed to show the interest of a historical approach to imagine the structuring of the French coast differently. Indeed, vulnerability in these regions increased considerably due to the choices made when arranging these territories. However, for centuries, societies have know to adapt to littoral risk by developing a culture of survival based on the precautionary principle. Paradoxically, it is modern society resulting from the industrial revolution which caused us to break away from this sustainable development model. It was thought that our powerful urban society, with the technical help of the engineers, was able to build a society free from maritime risk. From this point of view, the Xynthia catastrophe brutally revealed the limits of a speculative model which disregarded nature and its constraints.

6 Records of the trial, p. 55.

Paradoxically, the question of climate change in France is often used by the developers. After each new catastrophe, climate change is often used as an excuse to cover up the shortcomings of policymakers and their poor choices of development. In these circumstances, how could they be responsible for having built a building on the coast now struck by a sea surge, when climatologists warned that these storms were a recent evolution (less than 30 years)? And if these same climatologists and modelers promised to calculate the height of the next sea surge using their models, why would they want to change their development strategy after a disaster?

De facto, since 2010, we have noted that the solution which emerged was only a technical one. It was mainly the construction of new sea walls, re-calibrated according to scientific models. This way, urbanization close to the sea can continue due to new works conceived by engineers without needing to consider the withdrawal of populations to higher ground. This choice is made today by the majority of the French coastal towns to increase their resilience.

Figure 3.25. *Stele set up by the association of the victims of La Faute-sur-Mer with a plate on which the names and first names of the Xynthia victims are engraved (source: E. Garnier)*

Conclusion

When reading this book, we may be puzzled by the high number of international organizations, national, international, regional and sometimes global conventions, which all deal with coastal risk management. Some of these programmes have been running for many years or even decades, mostly within the industrialized northern hemisphere. A fundamental difficulty for an efficient interface between research and policy arises from the fact that research and policy have different and varying agendas [ROO 11]. Where policy tends to focus the short-term perspective, science envisages a long-term perspective. Moreover, while policy tries to involve the development of acceptable compromises, the scientific community aims to work toward the collection of objective scientific facts and the development of reasonable theories. This ambiguity is also present in the current monitoring programmes.

Our conception and perception of risks continue to evolve, notably under the pressure of multiple and multidimensional risks generated by climate change. Nevertheless, there is no emerging concertation at global level about the process of building up and transmission of a "culture of risk", which would allow us to prepare our societies to sometimes imminent threats including social, economic, environmental and political consequences. It appears that the decision-making process and prevention and management procedures of coastal risks, either at national, international or European levels, remain unable to take into account all spatial and temporal dimensions of threatening natural hazards while at the same time considering ecological and societal inequities, which also threaten the stability of our societies.

Vulnerability to risks may vary in relation to several factors and individual or societal group-level considerations. In this context, tools and instruments included in the ICZM framework enable us to improve traditional decision-making processes, closely associating citizens to decisions of concern to them for the prevention and management but also the acceptance of coastal risks. Environmental policies are integrated into the ICZM objectives and can support its implementation. It is, however, facing a lack of integration and coordination of different instruments at implementation levels.

Recent years have brought us many tools and techniques for assessing coastal risks related to storms, e.g. observing technologies such as lasers or airborne radars, satellite imaging, ground-based video surveillance systems, etc., which are complemented by field data. Due to the state-of-the-art technologies, as well as new numerical models and formulas, incorporating morphological characteristics of the beach and dunes, a better understanding of the influence of storm surges on coastal systems is now possible. These considerations are essential if we are to accurately assess the current and future risks to which coastal populations are subjected. Historical data on past storms are also very important as they allow time series of weather data to be constructed for predicting the characteristics of future storms. Climate models using different scenarios for occurrence of extreme events and sea-level rise complement the forecasts and determine the real challenges for coastal risk management in the long term. In addition to physical factors, socio-economic and environmental factors should be an integral part of assessing coastal storm and flood risk. In this context, an accurate assessment of potential losses or real losses of goods and services (natural and human, tangible and intangible) following an extreme storm event, it is necessary to estimate the vulnerability of coastal dwellings. As a final remark, it should be noted that although research has made significant progress in this area, storm risk assessment remains a major challenge and a crucial step in risk management of coastal areas of the European Union and beyond.

Bibliography

[ALE 04] ALEXANDER P.S., HOLMAN R.A., "Quantification of nearshore morphology based on video imaging", *Marine Geology*, vol. 208, pp. 101–111, 2004.

[ALV 10] ALVAREZ-ELLAcuria A., ORFILA A., OLABARRIETA M. *et al.*, "Nearshore Wave and Current Operational Forecasting System", *Journal of Coastal Research*, vol. 263, pp. 503–509, 2010.

[AND 07] ANDRADE C., PIRES H.O., TABORDA R. *et al.*, "Projecting future changes in wave climate and coastal response in Portugal by the end of the 21st century", *Journal of Coastal Research*, vol. SI 50, pp. 263–257, 2007.

[AND 13] ANDRÉ C., Analyse des dommages liés aux submersions marines et évaluation des coûts induits aux habitations à partir de données d'assurance. Perspectives apportées par les tempêtes Johanna (2008) et Xynthia (2010), Brest, PhD Thesis, University of Brittany, 2013.

[ANT 11] ANTUNES C., "Monitoring sea level change at Cascais tide gauge", *Journal of Coastal Research,* vol. SI 64, pp. 870–874, 2011.

[ARM 06] ARMAROLI C., CIAVOLA P., BALOUIN Y. *et al.*, "An Integrated Study of Shoreline Variability Using GIS and ARGUS Techniques", *Journal of Coastal Research*, vol. SI 39, pp. 1996–2000, 2006.

[ARM 09] ARMAROLI C., CIAVOLA P., MASINA M. *et al.*, "Run-up computation behind emerged breakwaters for marine storm risk assessment", *Journal of Coastal Research*, vol. SI 56, pp. 1612–1616, 2009.

[ART 09] ARTIGAS F., CHUN S.A., SOOKHU Y., "Real-time ocean surge warning system, meadowlands district of New Jersey", *10th Annual International Conference on Digital Government Research (Dg.o 2009)*, pp. 216–222, 2009.

[AYR 11] AYRAL P.A., GONZALEZ C., LEQUETTE C. *et al.*, *Workshop of The International Emergency Management Society,* France, June 2011.

[BAA 09] BAART F., KAAIJ T., VAN DER ORMONDT M. *et al.*, "Real-time forecasting of morphological storm impacts: A case study in the Netherlands", *Journal of Coastal Research*, vol. 2009, pp. 1617–1621, 2009.

[BAA 11] BAART F., BAKKER M.A.J., DONGEREN A. *et al.*, "Using 18th century storm-surge data from the Dutch Coast to improve the confidence in flood-risk estimates", *Natural Hazards and Earth System Sciences*, vol. 11, pp. 2791–2801, 2011.

[BAJ 07] BAJO M., ZAMPATO L., UMGIESSER G. *et al.*, *Estuarine, Coastal and Shelf Sciences*, vol. 75, pp. 236–249, 2007.

[BAJ 10] BAJO M., UMGIESSER G., "Storm surge forecast through a combination of dynamic and neural network models", *Ocean Modelling*, vol. 33, pp. 1–9, 2010.

[BAR 02] BARGAGLI A., CARILLO A., PISACANE G. *et al.*, "An Integrated Forecast System over the Mediterranean Basin: Extreme Surge Prediction in the Northern Adriatic Sea", *Monthly Weather Reviews*, vol. 130, pp. 1317–1332, 2002.

[BAR 08a] BARREDO J.I., SALAMON P., BÓDIS K., "Towards an assessment of coastal flood damage potential in Europe", JRC Scientific and Technical Report (n. EUR 23698 EN), 2008.

[BAR 08b] BARTHOLMES J.C., THIELEN J., RAMOS M.H. *et al.*, *Hydrological Earth System Science Discussions*, vol. 5, pp. 289–322, 2008.

[BAR 15] BARNARD P.L., SHORT A.D., HARLEY M.D. *et al.*, "The European Flood Alert System EFAS – Part 2: Statistical skill assessment of probabilistic and deterministic operational forecasts", *Nature Geosciences*, vol. 8, pp. 801–807, 2015.

[BAT 71] BATTJES J., "Run-Up Distributions of Waves Breaking on Slopes", *Journal of Waterways and Harbour Division*, vol. 97, pp. 91–114, 1971.

[BAT 08] BATES B.C., KUNDZEWICZ Z.W., WU S. *et al.* (eds), Climate Change and Water, Technical Paper of the Intergovernmental Panel on Climate Change, IPCC Secretariat, Geneva, p. 210, 2008.

[BER 12a] BERTIN X., BRUNEAU N., BREILH J.-F. *et al.*, "Importance of wave age and resonance in storm surges: the case Xynthia, Bay of Biscay", *Ocean Modelling*, vol. 42, pp. 16–30, 2012.

[BER 16] BERGSMA E.W.J., CONLEY D.C., DAVIDSON M.A. *et al.*, "Video-based nearshore bathymetry estimation in macro-tidal environments", *Marine Geology*, vol. 374, pp. 31–41, 2016.

[BOO 99] BOOIJ N., RIS R.C., HOLTHUIJSEN L.H., "A third-generation wave model for coastal regions: 1. Model description and validation", *Journal of Geophysical Research*, vol. 104, pp. 7649–7666, 1999.

[BOW 68] BOWEN A.J., INMAN D.L., SIMMONS V.P., "Wave set-down and set-up", *Journal of Geophysical Research*, vol. 73, pp. 2569–2577, 1968.

[BRO 09a] BROCK J.C., PURKIS S.J., "The emerging role of Lidar remote sensing in coastal research and resource management", *Journal of Coastal Research*, vol. SI 53, pp. 1–5, 2009.

[BRO 09b] BROWN J.M., WOLF J., "Coupled wave and surge modelling for the eastern Irish Sea and implications for model wind-stress", *Continental Shelf Research*, vol. 29, pp. 1329–1342, 2009.

[BRO 10] BROWN J.M., "A case study of combined wave and water levels under storm conditions using WAM and SWAN in a shallow water application", *Ocean Modelling*, vol. 35, pp. 215–229, 2010.

[BUC 05] BUCH E., SHE J., "Operational ocean forecasting at the Danish Meteorological Institute", *Environmental Research, Engineering and Management*, vol. 3, no. 33, pp. 5–11, 2005.

[BUG 13] BUGAJNY N., FURMAŃCZYK K., DUDZIŃSKA-NOWAK J. *et al.*, "Modelling morphological changes of beach and dune induced by storm on the Soutern Baltic coast using XBeach (case study: Dziwnow Spit)", *Journal of Coastal Research*, vol. SI 65, pp. 672–677, 2013.

[BUG 15] BUGAJNY N., FURMAŃCZYK K., DUDZIŃSKA-NOWAK J., "Application of XBeach to model a storm response on a sandy spit at the southern Baltic", *Oceanol. Hydrobiological Studies*, vol. 44, pp. 552–562, 2015.

[BUR 10] BURZEL A., DASSANAYAKE D., NAULIN M. *et al.*, "Integrated flood risk analysis for extreme storm surges (XTREMERISK)", *International Conference of Coastal Engineering*, ASCE, 2010.

[CAL 10] CALDWELL A.W., CONRADS P.A., MASON R.R. *et al.*, "USGS hurricane storm-surge monitoring networks: An example from Hurricane Rita", *Proceedings of the 2010 South Carolina Water Resources Conference*, 2010.

[CAN 14] CANS C., DINIZ I., PONTIER J.-M. *et al.* (eds), *Traité de droit des risques naturels*, Le Moniteur, Paris, 2014.

[CAR 00] CARRETERO ALBIACH J.C., ALVAREZ FANJUL E., GÓMEZ LAHOZ M. *et al.*, "Ocean forecasting in narrow shelf seas: Application to the Spanish coasts", *Coastal Engineering*, vol. 41, pp. 269–293, 2000.

[CIA 07] CIAVOLA P., ARMAROLI C., CHIGGIATO J. *et al.*, "Impact of storms along the coastline of Emilia-Romagna: the morphological signature on the Ravenna coastline (Italy)", *Journal of Coastal Research*, vol. SI 50, pp. 540–544, 2007.

[CIA 11a] CIAVOLA P., FERREIRA O., HAERENS P. *et al.*, "Storm impacts along European coastlines. Part 1: The joint effort of the MICORE and ConHaz Projects", *Environmental Science and Policy*, vol. 14, pp. 912–923, 2011.

[CIA 11b] CIAVOLA P., FERREIRA O., HAERENS P. *et al.*, "Storm impacts along European coastlines. Part 2: Lessons learned from the MICORE project", *Environmental Science and Policy*, vol. 14, pp. 924–933, 2011.

[CIA 12] CIAVOLA P., STIVE M.J.F., "Thresholds for storm impacts along European coastlines: introduction", *Geomorphology*, vol. 143–144, pp. 1–2, 2012.

[CIA 13] CIAVOLA P., JIMÉNEZ J.A., "Preface: the record of marine storminess along European coastlines", *Natural Hazards and Earth System Sciences*, vol. 13, pp. 1999–2002, 2013.

[CIE 08] CIEŚLIKIEWICZ W., PAPLIŃSKA-SWERPEL B., "A 44-year hindcast of wind wave fields over the Baltic Sea", *Coastal Engineering*, vol. 55, pp. 894–905, 2008.

[CLA 99] CLAESSENS E.J., WAAL J.P., De TWUIVER H.C. *et al.*, "The Application of a Coupled Wind, Water Level and Wave Model in a Warning System Against Flooding", *Seventh International Workshop on Wave Hindcasting and Forecasting*, Banff, 1999.

[DAW 11] DAWSON R.J., PEPPE R., WANG M., "An agent-based model for risk-based flood incident management", *Natural Hazards*, vol. 59, pp. 167–189, 2011.

[DIN 09] DING Y., WANG S.S.Y., "Development of a Dynamic Data Driven Coastal / Estuarine Process Modeling System for Application to Coastal Water Environmental Prediction", *33rd IAHR Congress: Water Engineering for a Sustainable Environment* (IAHR), pp. 2083–2090, 2009.

[DON 06] DONNELLY C., KRAUS N., LARSON M., "State of Knowledge on Measurement and Modeling of Coastal Overwash" , *Journal of Coastal Research*, vol. 224, pp. 965–991, 2006.

[DON 08] DONGEREN A.V., PLANT N., COHEN A. *et al.*, "Beach Wizard: Nearshore bathymetry estimation through assimilation of model computations and remote observations", *Coastal Engineering*, vol. 55, pp. 1016–1027, 2008.

[DON 14] DONGEREN A.V., CIAVOLA P., VIAVATTENE C. *et al.*, "RISC-KIT: Resilience-Increasing Strategies for Coasts – ToolKIT", *Journal of Coastal Research*, vol. 70, 2014.

[DUO 15] DUO E., CIAVOLA P. (eds), Review report of key challenges and lessons learned from historical extreme hydro-meteorological events, Risc-Kit Project, Deliverable no. D.1.1, 2015.

[EFF 10] EFFSA La tempête Xynthia du 28 février 2010 – Bilan chiffré, Fédération Française des Sociétés d'Assurances-GEMA Assureurs Mutualistes (www.effsa.fr), 2010.

[ENV 11] ENVIRONMENT AGENCY, UK Coastal Monitoring and Forecasting (UKCMF) Service : Service Definition Bristol, 2011.

[EST 12] ESTEVES L.S., BROWN J.M., WILLIAMS J.J. *et al.*, "Quantifying thresholds for significant dune erosion along the Sefton Coast, Northwest England", *Geomorphology*, vol. 143–144, pp. 52–61, 2012.

[EUR 94] EUROPEAN COMMISSION, *Official Journal of the European Commission*, vol. C 135, 8 May 1994.

[EUR 00] EUROPEAN COMMISSION, "Directive 2000/60/EC of the European Parliament and of the Council of 23 October 2000 establishing a framework for Community action in the field of water policy", Official Journal of the European Commission, vol. L 327, p. 1, 2000.

[EUR 02] EUROPEAN COMMISSION, COM/00/547 of 17 September 2002, Official Journal of the European Commission, no. 148, 6 June 2002.

[EUR 06] EUROPEAN COMMISSION, Green paper – towards a EU maritime policy, Official Journal of the European Commission, COM(2006) 275 final, 2006.

[EUR 07] EUROPEAN COMMISSION, Directive 2007/60/EC of the European Parliament and of the Council of 23 October 2007 on the assessment and management of flood risks, Official Journal of the European Commission, vol. L 288, p. 27, 2007.

[EUR 08] EUROPEAN COMMISSION, EU Marine Strategy Framework Directive, 2008/56/EC, Official Journal of the European Commission, 2008.

[EUR 09a] EUROPEAN COMMISSION, Towards a European Integrated Coastal Zone Management (ICZM) strategy: general principles and policy options, reflection paper, available at: http://ec.europa.eu/environment/iczm/pdf/vol1.pdf, 2009.

[EUR 09b] EUROPEAN COMMISSION, European Research Framework Programme. Research on climate change, European Commission, EUR 23609, 2009.

[EUR 09c] EUROPEAN COMMISSION, Madrid Protocol signed on 4 December 2008, Official Journal of the European Commission, vol. L 034, 4 February 2009.

[EUR 10a] EUROPEAN COMMISSION, Internal Security Strategy for the European Union: towards a European Security Model, 5842/2/2010, 2010.

[EUR 10b] EUROPEAN ENVIRONMENT AGENCY, Mapping the impacts of recent natural disasters and technological accidents in Europe – an overview of the last decade, 2010.

[EUR 11] EUROPEAN COMMISSION, "Climate Change Impacts and Adaptation: reducing water-related risks in Europe", *Proceedings of the EU-ISDR Int. Workshop*, Brussels, 6–7 July 2010, EUR Report, EUR 10-620 EN, 2011.

[EUR 12a] EUROPEAN COMMISSION, EU Industrial Policy, COM 417 final, 2012.

[EUR 12b] EUROPEAN COMMISSION, COM 2012/0337, 2012.

[EUR 13a] EUROPEAN COMMISSION, EU Civil Protection Mechanism, Decision 1313/2013, 2013.

[EUR 13b] EUROPEAN COMMISSION, Decision 1082/2013, 2013.

[EUR 13c] EUROPEAN COMMISSION, COM 216 final, 2013.

[EUR 13d] EUROPEAN COMMISSION, SWD 318 final, 2013.

[EUR 14] EUROPEAN ENVIRONMENT, Agency, National adaptation policy processes in European countries, 2014.

[EUR 15] EUROPEAN COMMISSION, The European Agenda on Security, COM(2015) 185 final, 2015.

[EWA 86] EWALD E., *L'Etat-providence*, Grasset, Paris, 1986.

[FAN 01] FANJUL E.Á., GÓMEZ B.P., ARÉVALO I.R.S., "Nivmar: a storm surge forecasting system for Spanish waters", *Scientia Marina*, vol. 65, pp. 145–154, 2001.

[FEM 03] FEMA, Guidelines and Specifications for Flood Hazard Mapping Partners Report of the Federal Emergency Management Agency, US Government, 2003.

[FEN 08] FENGER J., BUCH E., JAKOBSEN P.R. *et al.*, "Danish Attitudes and Reactions to the Threat of Sea-Level Rise", *Journal of Coastal Research*, vol. 242, pp. 394–402, 2008.

[FIN 05] FINKL C.W., BENEDET L., ANDREWS J.L., "Submarine Geomorphology of the Continental Shelf off Southeast Florida Based on Interpretation of Airborne Laser Bathymetry", *Journal of Coastal Research*, vol. 216, pp. 1178–1190, 2005.

[FLA 00] FLATHER R.A., "Existing operational oceanography", *Coastal Engineering*, vol. 41, pp. 13–40, 2000.

[FLO 10] FLOWERDEW J., HORSBURGHB K., WILSON C. *et al.* "Development and evaluation of an ensemble forecasting system for coastal storm surges", *Quarterly Journal of the Royal Meteorological Society*, vol. 136, pp. 1444–1456, 2010.

[GAL 14] GALLIEN T.W., SANDERS B.F., FLICK R.E., "Urban coastal flood prediction: Integrating wave overtopping, flood defenses and drainage", *Coastal Engineering*, vol. 91, pp. 18–28, 2014.

[GAR 10a] GARNIER E. (ed.), La crise Xynthia à l'aune de l'histoire aux missions d'enquêtes parlementaire et sénatoriale sur Xynthia, Report for the French Parliament Inquiry, 2010.

[GAR 10b] GARNIER E., *Les dérangements du temps, 500 ans de chaud et froids en Europe,* Plon, Paris, 2010.

[GAR 11] GARNIER E., SURVILLE F. (eds), *La tempête Xynthia face à l'histoire. Submersions et tsunamis sur les littoraux français du Moyen Age à nos jours*, Le Croît vif, Saintes, 2011.

[GAR 12] GARNIER E., HENRY N., DESARTHE J., "Visions croisées de l'historien et du courtier en réassurance sur les submersions. Recrudescence de l'aléa ou vulnérabilisation croissante ?", in HALLEGATTE S., PRZYLUSKI V. (eds), *Gestion des risques naturels. Leçons de la tempête Xynthia*, Editions Quae, Paris, 2012.

[GAR 15a] GARNIER E., "A historic experience for a strengthened resilience. European societies in front of hydro-meteors 16th–20th centuries", in QUEVAUVILLER P. (ed.), *Prevention of Hydrometeorological Extreme Events-Interfacing Sciences and Policies*, John Wiley & Sons, Chichester, 2015.

[GAR 15b] GARNIER E., "Influence des actions anthropiques", *Développer la connaissance et l'observation du trait de côte. Contribution nationale pour une gestion intégrée pour la COP 21*, Ministère de l'Ecologie, du Développement durable et de l'Energie, Paris, pp. 7–10, 2015.

[GLA 09] GLAHN B., TAYLOR A., KURKOWSKI N. *et al.*, "The role of the SLOSH model in National Weather Service storm surge forecasting", *National Weather Digest*, pp. 1–12, 2009.

[GUT 15] GUTIERREZ B.T., PLANT N.G., THIELER E.R. *et al.*, "Using a Bayesian network to predict barrier island geomorphologic characteristics", *Journal of Geophysical Research F Earth Surf.*, vol. 120, pp. 2452–2475, 2015.

[GUZ 81] GUZA R.T., THORNTON E.B., "Wave set-up on a natural beach", *Journal of Geophysical Research*, vol. 86, pp. 4133–4137, 1981.

[HAL 08] HALLEGATTE S., "An adaptive regional input-output model and its application to the assessment of the economic cost of Katrina", *Risk Analysis* vol. 28, pp. 779–799, 2008.

[HAN 93] HANSLOW D., NIELSEN P., "Shoreline set-up on natural beaches", *Journal of Coastal Research*, vol. SI 15, pp. 1–10, 1993.

[HAR 12] HARLEY M.D., VALENTINI A., ARMAROLI C. *et al.*, "An early warning system for the on-line prediction of coastal storm risk on the Italian coastline, *International Conference of Coastal Engineering* ASCE, 2012.

[HAR 13] HARLEY M.D., CIAVOLA P., "Managing local coastal inundation risk using real-time forecasts and artificial dune placements", *Coastal Engineering*, vol. 77, pp. 77–90, 2013.

[HAR 14] HARLEY M.D., ANDRIOLO U., ARMAROLI C. *et al.*, "Shoreline rotation and response to nourishment of a gravel embayed beach using a low-cost video monitoring technique: San Michele-Sassi Neri", *Journal of Coastal Conservation*, vol. 18, 2014.

[HAR 16] HARLEY M.D., VALENTINI A., ARMAROLI C. *et al.*, "Can an early-warning system help minimize the impacts of coastal storms? A case study of the 2012 Halloween storm, northern Italy, Central Italy", *Natural Hazards and Earth System Sciences*, vol. 16, pp. 209–222, 2016.

[HEU 08] HEURTEFEUX H., LESAIGNOUX A., DENAMIEL C., "Assessment, validation and further uses of LiDAR survey in the Western part of French Mediterranean sea", in PRANZINI E., WETZEL L. (eds), *Beach Erosion Monitoring*, La nuova Grafica Fiorentina, Florence, Italy, 2008.

[HIN 10] HINKEL J., NICHOLLS R.J., VAFEIDIS A.T. *et al.*, "Assessing risk of and adaptation to sea-level rise in the European Union: An application of DIVA", *Mitigation and Adaptation Strategies for Global Change*, vol. 15, pp. 703–719, 2010.

[HOL 85] HOLMAN R.A., SALLENGER A.H., "Setup and swash on a natural beach", *Journal of Geophysical Research*, vol. 90, p. 945, 1985.

[HOL 86] HOLMAN R.A., "Extreme value statistics for wave run-up on a natural beach", *Coastal Engineering*, vol. 9, pp. 527–544, 1986.

[HOL 07] HOLMAN R.A., Stanley J., "The history and technical capabilities of Argus", *Coastal Engineering*, vol. 54, pp. 477–491, 2007.

[HOL 13] HOLMAN R., PLANT N., HOLLAND T., "CBathy: A robust algorithm for estimating nearshore bathymetry", *Journal of Geophysical Research Ocean.*, vol. 118, pp. 2595–2609, 2013.

[HON 10] HONDULA D.M., DOLAN R., "Predicting severe winter coastal storm damage", *Environmental Research Letters*, vol. 5, p. 34004, 2010.

[HOR 09] HORSBURGH K.J., BRADLEY L., ANGUS M. *et al.*, High frequency sea level recording for tsunami warning and enhanced storm surge monitoring at UK sites, Report, Prooudman Oceanographic Laboratory, Natural Environment Research Council, Liverpool, 2009.

[HUN 59] HUNT I.A., "Design of seawalls and breakwaters", *Journal of Waterways and Harbours Division*, vol. 85, pp. 123–152, 1959.

[IGL 10] IGLESIAS-CAMPOS A., SIMON-COLINA A., FRAILE-JURADO P. *et al.*, Methods for assessing current and future coastal vulnerability to climate change, ETC/ACC Technical Paper 2010/8, 2010.

[IPC 14] IPCC, Climate change 2014: impacts, adaptation and vulnerability, Chapter V on Coastal Systems and Low-Lying Areas, p. 381, 2014.

[IRI 49] IRIBARREN C.R., NOGALES C.M, "Protection des ports", *PIANC Congress 1949 SII-C4*, pp. 180–193, 1949.

[IRI 98] IRISH J.L., WHITE T.E., "Coastal engineering applications of high-resolution lidar bathymetry", *Coastal Engineering*, vol. 35, pp. 47–71, 1998.

[KAY 05] KAYA Y., STEWART M., BECKER M., "Flood forecasting and flood warning in the Firth of Clyde, UK", *Natural Hazards*, vol. 36, pp. 257–271, 2005.

[KLE 09] KLEMAS V.V., "The Role of Remote Sensing in Predicting and Determining Coastal Storm Impacts", *Journal of Coastal Research*, vol. 256, pp. 1264–1275, 2009.

[KOL 09] KOLEN B., KUTSCHERA G., HELSOOT I., A Comparison between the Netherlands and Germany of Evacuation in Case of Extreme Flooding, *Urban Flood Conference,* Paris, 26–27 November, 2009.

[KOL 10] KOLEN B., SLOMP R., VAN BALEN W. *et al.*, Learning from French experiences with storm Xynthia. Dams after a flood, Report, Ministerie van Verkeer en Waterstaat, Arnhem, 2010.

[KOM 72] KOMAR P.D., GAUGHAN M.K., *Proceedings 13th Conference of Coastal Engineering, International Conference of Coastal Engineering*, pp. 405–418, 1972.

[KRE 14] KREIBICH H., BERGH J.C.J.M., VAN DEN BOUWER L.M. *et al.*, "Costing natural hazards", *Nature Climate Change*, vol. 4. 2014.

[LAG 11] LAGEMAA P., ELKEN J., KÕUTS T., "Operatiivne meretaseme prognoosisüsteem eestis", *Estonian Journal of Engineering*, vol. 17, pp. 301–331, 2011.

[LAM 91a] LAMB H., *Historical Storms of the North Sea, British Isles and Norhwest Europe*, Cambridge University Press, Cambridge, 1991.

[LAR 04] LARSON M., ERIKSON L., HANSON H., "An analytical model to predict dune erosion due to wave impact", *Coastal Engineering*, vol. 51, pp. 675–696, 2004.

[LAR 15] LARONDE-CLÉRAC C., MAZEAUD A., MICHELOT A., *Les risques naturels en zones côtières – Xynthia: enjeux politiques, questionnements juridiques*, Presses Universitaires de Rennes, 2015.

[LEE 11] LEE C.B., KAO C.C., DOONG D.J. *et al.*, "Typhoon Data Measurements and Quality Control", 8th *OMISAR Workshop on Ocean Models*, 2011.

[LIB 13] LIBERATO M., PINTO J., TRIGO R. *et al.*, "Explosive development of winter storm Xynthia over the subtropical North Atlantic Ocean", *Natural Hazards and Earth System Science*, vol. 13, pp. 2239–2251, 2013.

[LIF 96] LIFE, Responding to risks linked to climate change in coastal Life-Environment 2003–2006, RESPONSE project, zones, 1996.

[LIO 05] LIONELLO P., SANNA A., "Mediterranean wave climate variability and its links with NAO and Indian Monsoon", *Climate Dynamics*, vol. 25, pp. 611–623, 2005.

[LIO 06] LIONELLO P., SANNA A., ELVINI E. *et al.*, "A data assimilation procedure for operational prediction of storm surge in the northern Adriatic Sea", *Continental Shelf Research*, vol. 26, pp. 539–553, 2006.

[LUM 11] LUMBROSO D.M., VINET F., "A comparison of the causes, effects and aftermaths of the coastal flooding of England in 1953 and France in 2010", *Natural Hazards and Earth System Sciences*, vol. 11, pp. 2321–2333, 2011.

[MAS 00] MASON D.C., GURNEY C., KENNETT M., "Beach Topography Mapping: a Comparison of Techniques", *Journal of Coastal Conservation*, vol. 6, pp. 113–124, 2000.

[MAS 15] MAS E., BRICKER J., KURE S. *et al.*, "Field survey report and satellite image interpretation of the 2013 Super Typhoon Haiyan in the Philippines", *Natural Hazards and Earth System Sciences*, vol. 15, pp. 805–816, 2015.

[MCI 06] MCINTYRE M.L., NAAR D.F., CARDER K.L. *et al.*, "Coastal Bathymetry from Hyperspectral Remote Sensing Data: Comparisons with High Resolution Multibeam Bathymetry", *Marine Geophysical Researches*, vol. 27, pp. 129–136, 2006.

[MEL 10] MELTON G., GALL M., MITCHELL J.T. *et al.*, "Hurricane Katrina storm surge delineation: Implications for future storm surge forecasts and warnings", *Natural Hazards*, vol. 54, pp. 519–536, 2010.

[MIT 03] MITASOVA H., BERNSTEIN D.J., DRAKE T.G. *et al.*, "Spatio-Temporal Analysis of Beach Morphology using LiDAR, RTK-GPS and Open Source GRASS GIS", *International Conference on Coastal Sediments*, 2003.

[NIC 11] NICHOLLS R.J., "Planning for the impacts of sea level rise", *Oceanography*, vol. 24, pp. 144–157, 2011.

[NIE 91] NIELSEN P., HANSLOW D.J., "Wave Runup Distributions on Natural Beaches", *Journal of Coastal Research*, vol. 7, pp. 1139–1152, 1991.

[PAC 15] PACHECO A., HORTA J., LOUREIRO C. *et al.*, "Retrieval of nearshore bathymetry from Landsat 8 images: a tool for coastal monitoring in shallow waters", *Remote Sensing of the Environment*, vol. 159, pp. 102–116, 2015.

[PAU 09] PAUL B.K., "Why relatively fewer people died? the case of Bangladesh's cyclone SIDR", *Natural Hazards*, vol. 50, pp. 289–304, 2009.

[PER 16] PERINI L., CALABRESE L., SALERNO G. *et al.*, "Evaluation of coastal vulnerability to flooding: Comparison of two different methodologies adopted by the Emilia-Romagna region (Italy)", *Natural Hazards and Earth System Sciences*, vol. 16, 2016.

[PFI 10] PFISTER C., GARNIER E., ALCOFORADO M.J. *et al.*, "The meteorological framework and the cultural memory of three severe winter-storms in early eighteenth-century Europe", *Climate Change*, vol. 101, pp. 281–310, 2010.

[PIN 10] PINEAU-GUILLOU L., LATHUILLIERE C., MAGNE R. *et al.*, "Caractérisation des niveaux marins et modélisation des surcotes pendant le tempête Xynthia", *XIe Journées Nationales Génie Côtier-Génie Civils*, 2010.

[PRO 83] PROCTOR R., FLATHER R.A., Routine storm surge forecasting using numerical models: procedures and computer programs for use on the CDC Cyber 205E at the British Meteorological Office, Report, Institute of Oceanographic Sciences, 1983.

[QUE 10] QUEVAUVILLER PH. (ed.), *Water System Science and Policy Interfacing*, RSC Publishing, Cambridge, 2010.

[QUE 11a] QUEVAUVILLER PH., "Adapting to climate change: reducing water-related risks in Europe – EU policy and research considerations", *Environmental Science and Policy*, vol. 14, no. 7, p. 722, 2011.

[QUE 11b] QUEVAUVILLER PH., BORCHERS U., THOMPSON K.C. *et al.* (eds), *The Water Framework Directive – Action Programmes and Adaptation to Climate Change*, RSC Publishing, Cambridge, 2011.

[QUE 11c] QUEVAUVILLER PH., ROOSE P., VERREET G. (eds), *Chemical Marine Monitoring – Policy Framework and Analytical Trends*, John Wiley & Sons., Chichester, 2011.

[QUE 14] QUEVAUVILLER PH. (ed.), *Hydrometeorological Hazards – Interfacing Science and Policy, Hydrometeorological Extreme Events Series*, John Wiley & Sons Ltd., Chichester, 2014.

[RAP 11] RAPOSEIRO P.D., FERREIRA J.C., "Evaluation of coastal flood risk areas and adaptation strategies for a sustainable planning", *Journal of Coastal Research*, pp. 1896–1900, 2011.

[ROE 09] ROELVINK D., RENIERS A., VAN DONGEREN A. *et al.*, "Modelling storm impacts on beaches, dunes and barrier islands", *Coastal Engineering*, vol. 56, pp. 1133–1152, 2009.

[ROO 11] ROO A. DE, THIELEN J., SALAMON P. *et al.*, "Quality control, validation and user feedback of the European Flood Alert System (EFAS)", *International Journal of Digital Earth*, vol. 4, pp. 77–90, 2011.

[RUE 02] RUESSINK B.G., BELL P.S., ENCKEVORT I.M.J. *et al.*, "Nearshore bar crest location quantified from time-averaged X-band radar images", *Coastal Engineering*, vol. 45, pp. 19–32, 2002.

[RUG 01] RUGGIERO P., KOMAR P.D., MCDOUGAL W.G. *et al.*, "Wave runup, extreme water levels and the erosion of properties backing beaches", *Journal of Coastal Research*, vol. 17, pp. 407–419, 2001.

[RUG 04] RUGGIERO P., HOLMAN R.A., BEACH R.A., "Wave run-up on a high-energy dissipative beach", *Journal of Geophysical Research, C. Ocean.*, vol. 109, 2004.

[SAL 00] SALLENGER A.H., "Storm Impact Scale for Barrier Islands", *Journal of Coastal Research*, vol. 16, pp. 890–895, 2000.

[SAL 03] SALLENGER A.H., HOWD P., STOCKDON H. *et al.*, "On predicting storm-induced coastal change", *International Conference on Coastal Sediments*, pp. 1–9, 2003.

[SCA 06] SCAWTHORN C., FLORES P., BLAIS N. *et al.*, "HAZUS-MH Flood Loss Estimation Methodology. II. Damage and Loss Assessment", *Natural Hazards Review Rev.*, vol. 7, pp. 72–81, 2006.

[SCH 06] SCHNEIDER P.J., SCHAUER B., "HAZUS—Its Development and Its Future", *Natural Hazards Review*, vol. 7, pp. 40–44, 2006.

[SOT 06] SOTILLO M.G., AZNAR R., VALERO F., "Mediterranean offshore extreme wind analysis from the 44-year HIPOCAS database: different approaches towards the estimation of return periods and levels of extreme values", *Advances in Geosciences*, vol. 7, pp. 275–278, 2006.

[SPE 16] SPENCER T., SCHUERCH M., NICHOLLS R.J. *et al.*, "Global coastal wetland change under sea-level rise and related stresses: The DIVA Wetland Change Model", *Glob. Planet. Change*. 139, pp. 15–30, 2016.

[STO 06a] STOCKDON H.F., HOLMAN R.A., HOWD P.A. *et al.*, "Empirical parameterization of setup, swash, and runup", *Coastal Engineering*, vol. 53, pp. 573–588, 2006.

[STO 06b] STORCH VON H., WOTH K., "Storm surges – the case of Hamburg, Germany, GEC, natural disasters, and their implications for human security in coastal urban areas", *ESSP OSC panel session*, p. 5, 2006.

[THI 09] THIELEN J., BARTHOLMES J., RAMOS M.H. *et al.*, "The European Flood Alert System – Part 1: Concept and development", *Hydrology and Earth System Sciences*, vol. 13, pp. 125–140, 2009.

[TOL 13] TOLLEFSON J., "Hurricane Sandy spins up climate discussion, *Nature News & Comments*, pp. 17–19, 2013.

[TUR 16] TURNER I.L., HARLEY M.D., DRUMMOND C.D., "UAVs for coastal surveying", *Coastal Engineering*, vol. 114, pp. 19–24, 2016.

[UNI 82] UNITED NATIONS, UN Convention on the Law of the Sea (UNCLOS), available at: http://www.un.org/Depts/los/convention_agreements/texts/unclos/closindx.htm, 1982.

[UNI 92] UNITED NATIONS, Conference on Environment and Development available at: http://www.un.org/geninfo/bp/enviro.html, 1992.

[UNI 09] UNITED NATIONS, Guidance on Water and Adaptation to Climate Change, 2009.

[UNI 10] UNITED NATIONS, International Strategy for Disaster Reduction (ISDR), 2010–2011 Biennial Work Programme, Geneva, 2010.

[UNI 15] UNITED NATIONS, Sendai Framework for Action, 2015.

[VER 05] VERLAAN M., ZIJDERVELD A., DE VRIES H. *et al.*, "Operational storm surge forecasting in the Netherlands: developments in the last decade", *Philosophical transactions. Series A, mathematical, physical, and engineering sciences*, vol. 363, pp. 1441–1453, 2005.

[VIC 06a] VICKERY P.J., LIN J., SKERLJ P.F. *et al.*, "HAZUS-MH Hurricane Model Methodology. I: Hurricane Hazard, Terrain, and Wind Load Modeling", *Natural Hazards Review*, vol. 7, pp. 82–93, 2006.

[VIC 06b] VICKER P.J., SKERLJ P.F., LIN J. *et al.*, "HAZUS-MH Hurricane Model Methodology. II: Damage and Loss Estimation", *Natural Hazards Review*, vol. 7, pp. 94–103, 2006.

[VIN 12] VINET F., LUMBROSO D., DEFOSSEZ S. *et al.*, "A comparative analysis of the loss of life during two recent floods in France: the sea surge caused by the storm Xynthia and the flash flood in Var", *Natural Hazards*, vol. 61, pp. 1179–1201, 2012.

[VOU 12] VOUSDOUKAS M.I., WZIATEK D., ALMEIDA L.P., "Coastal vulnerability assessment based on video wave run-up observations at a mesotidal, steep-sloped beach", *Ocean Dynamics*, vol. 62, pp. 123–137, 2012.

[VOU 16] VOUSDOUKAS M.I., VOUKOUVALAS E., MENTASCHI L. *et al.*, "Developments in large-scale coastal flood hazard mapping", *Natural Hazards and Earth System Sciences*, vol. 16, pp. 1841–1853, 2016.

[VRI 95] DE VRIES H., BRETON M., MULDER T. de *et al.*, "A comparison of 2D storm surge models applied to three shallow European seas", *Environmental Software*, vol. 10, pp. 23–42, 1995.

[WES 01] WEST J.J., SMALL M.J., DOWLATABADI H., "Storms, investor decisions, and the economic impacts of sea level rise", *Climatic Change*, vol. 48, pp. 317–342, 2001.

[WOL 09] WOLF J., "Coastal flooding: Impacts of coupled wave-surge-tide models", *Natural Hazards*, vol. 49, pp. 241–260, 2009.

[WRI 84] WRIGHT L.D., SHORT A.D., "Morphodynamic variability of surf zones and beaches: a synthesis", *Marine Geology*, vol. 56, pp. 93–118, 1984.

Index

T, V

V, W, X

Other titles from

ISTE

in

Earth Systems – Environmental Engineering

2016

BLAIN Stéphane, TAGLIABUE Alessandro
Iron Cycle in Oceans

BRAUDEAU Erik, ASSI Amjad, MOHTAR Rabi
Hydrostructural Pedology

MONACO André, PROUZET Patrick
Seas and Oceans Set
Volume 9 – Tools for Oceanography and Ecosystemic Modeling

QUÉGUINER Bernard
The Biogeochemical Cycle of Silicon in the Ocean

QUEVAUVILLER Philippe
Marine Chemical Monitoring

2015

HERBIN Hervé, DUBUISSON Philippe
Infrared Observation of Earth's Atmosphere

MERCIER-LAURENT Eunika
The Innovation Biosphere

MONACO André, PROUZET Patrick
Seas and Oceans Set
Volume 7 – Marine Ecosystems: Diversity and Functions
Volume 8 – Governance of Seas and Oceans

2014

DE LARMINAT Philippe
Climate Change: Identification and Projections

MONACO André, PROUZET Patrick
Seas and Oceans Set
Volume 1 – Ocean in the Earth System
Volume 2 – The Land-Sea Interactions
Volume 3 – Ecosystem Sustainability and Global Change
Volume 4 – Vulnerability of Coastal Ecosystems and Adaptation
Volume 5 – Development of Marine Resources
Volume 6 – Value and Economy of Marine Resources

MUTTIN Frédéric
Marine Coastal and Water Pollutions: Oil Spill Studies

2013

LÉCUYER Christophe
Water on Earth: Physicochemical and Biological Properties

LEGCHENKO Anatoly
Magnetic Resonance Imaging for Groundwater

2012

LE MENN Marc
Instrumentation and Metrology in Oceanography

PAVÉ Alain
Modeling of Living Systems: From Cell to Ecosystem

2010

TANGUY Jean-Michel
Environmental Hydraulics Series
Volume 1 – Physical Processes and Measurement Devices
Volume 2 – Mathematical Models
Volume 3 – Numerical Methods
Volume 4 – Practical Applications in Engineering
Volume 5 – Modeling Software

Printed and bound by CPI Group (UK) Ltd, Croydon, CR0 4YY
13/09/2026
14961024-0001